The Road to B5G/6G Mobile Communication Networks: Technologies and Applications

RIVER PUBLISHERS SERIES IN COMMUNICATIONS AND NETWORKING

Series Editors:

ABBAS JAMALIPOUR
The University of Sydney, Australia

MARINA RUGGIERI
University of Rome Tor Vergata, Italy

MARKO JURCEVIC
University of Zagreb, Croatia

The "River Publishers Series in Communications and Networking" is a series of comprehensive academic and professional books which focus on communication and network systems. Topics range from the theory and use of systems involving all terminals, computers, and information processors to wired and wireless networks and network layouts, protocols, architectures, and implementations. Also covered are developments stemming from new market demands in systems, products, and technologies such as personal communications services, multimedia systems, enterprise networks, and optical communications.

The series includes research monographs, edited volumes, handbooks and textbooks, providing professionals, researchers, educators, and advanced students in the field with an invaluable insight into the latest research and developments.

Topics included in this series include:-

- Communication theory
- Multimedia systems
- Network architecture
- Optical communications
- Personal communication services
- Telecoms networks
- Wifi network protocols

For a list of other books in this series, visit www.riverpublishers.com

The Road to B5G/6G Mobile Communication Networks: Technologies and Applications

Editors

Chih-Cheng Tseng

National Taiwan Ocean University, Taiwan (ROC)

Albena Mihovska

SmartAvatar, Denmark

Shao-Yu Lien

National Yang Ming Chiao Tung University, Taiwan (ROC)

River Publishers

Routledge
Taylor & Francis Group
NEW YORK AND LONDON

Published 2025 by River Publishers
River Publishers
Broagervej 10, 9260 Gistrup, Denmark
www.riverpublishers.com

Distributed exclusively by Routledge
605 Third Avenue, New York, NY 10017, USA
4 Park Square, Milton Park, Abingdon, Oxon OX14 4RN

The Road to B5G/6G Mobile Communication Networks: Technologies and Applications / by Chih-Cheng Tseng, Albena Mihovska, Shao-Yu Lien.

Routledge is an imprint of the Taylor & Francis Group, an informa business

ISBN 978-87-4380-109-2 (hardback)
ISBN 978-87-4380-674-5 (paperback)
ISBN 978-87-4380-111-5 (online)
ISBN 978-87-4380-110-8 (master ebook)

DOI 10.1201/9788743801108

Contents

Preface

From the early days of analog voice transmission to today's high-speed digital platforms, wireless mobile communication has continually reshaped the way we live, work, and interact. As we enter the era of Beyond 5G (B5G) and sixth-generation (6G) networks, the demand for faster, smarter, and more energy-efficient connectivity becomes increasingly urgent. *The Road to B5G/6G Mobile Communication Networks: Technologies and Applications* provides a comprehensive exploration of this transition, spotlighting the innovations that are set to define the future of wireless systems.

More than just a technical overview, this book serves as a versatile reference for engineers, researchers, academics, and industry professionals. It offers a thorough treatment of the foundational principles, emerging technologies, and transformative applications that will drive the development of B5G and 6G networks.

While 5G has brought remarkable advances, it still faces challenges such as latency, limited scalability, and spectral inefficiencies. This book addresses these shortcomings by exploring a range of promising technologies, including Terahertz (THz) spectrum utilization, AI-native networking, ultra-massive MIMO, and reconfigurable intelligent surfaces (RIS). These technologies not only promise performance enhancements but also open new frontiers in sectors like healthcare, autonomous systems, smart cities, and immersive media.

The content is structured to guide readers from the key motivations behind B5G/6G, such as explosive data growth, limited spectrum resources, and the rise of edge intelligence, to the enabling technologies and architectural shifts required. Topics such as non-terrestrial networks (NTNs), machine learning for network optimization, and distributed, cloud-native infrastructures are examined in detail.

What sets this book apart is its focus on practical applications. Technological development is driven by real-world demands, and we delve into forward-looking scenarios such as the tactile internet, haptic feedback systems, holographic communication, and the integration of cyber-physical systems. These use cases are explored from technical, regulatory, and societal perspectives, ensuring a well-rounded analysis.

Equally emphasized is the importance of sustainability and social impact. As we develop future networks, considerations such as energy efficiency, environmental responsibility, and equitable access must remain central. This book integrates these values with technical insights, reflecting a broader vision of progress that benefits all.

The path to B5G and 6G is neither straightforward nor isolated. It is a collaborative, interdisciplinary endeavor involving researchers, industry leaders, policymakers, and end users. Our aim with this book is to provide an up-to-date, accessible guide that not only informs but also inspires continued innovation and exploration in the field of mobile communication.

We hope this work will serve as both a trusted reference and a source of inspiration for those helping to shape the future of global connectivity.

List of Figures

List of Tables

List of Contributors

Aldossari, Saud Alhajaj, *Department of Electrical Engineering, College of Engineering in Wadi Alddawasir, Prince Sattam Bin Abdulaziz University, Saudi Arabia*

Arslan, Hüseyin, *School of Engineering and Natural Sciences, Istanbul Medipol University, Turkey*

Cay, Buminhan, *Department of Research and Development, ULAK Communications Inc., Turkey*

Chen, Peng, *China Telecom Research Institute, China*

Das, Rajib K., *Department of Computer Science and Engineering, University of Calcutta, India*

Demircioglu, Erdem, *Department of Research and Development, ULAK Communications Inc., Turkey*

Gao, Lei, *China Telecom Research Institute, China*

Gkountara, Despoina N., *University of Huddersfield, United Kingdom*

Gupta, Vaibhav, *School of Electronics Engineering, Vellore Institute of Technology, India*

Hemanshi, Anil K., *School of Electronics Engineering, Vellore Institute of Technology, India*

Hussain, Ahmed A., *Department of Electrical Engineering, Prince Mohammad University, Saudi Arabia*

Ikenaga, T., *Department of Electrical and Electronic Engineering, Faculty of Engineering, Kyushu Institute of Technology, Japan*

Jaipuria, Saumya, *Department of Computer Science and Engineering, University of Calcutta, India*

Khara, Sibaram, *Sharda University, India*

Kim, Haesik, *VTT Technical Research Centre of Finland, Finland*

Kojima, F., *Social Innovation Unit, Open Innovation Platform Headquarters, National Institute of Information and Communications Technology (NICT), Japan*

Lazaridis, Pavlos I., *University of Huddersfield, United Kingdom*

Mayfield, Sebastian, *Department of Engineering & Technology, Texas A & M University, United States*

Miura, Amane, *National Institute of Information and Communications Technology, Japan*

Moinuddeen, AbuMuhammad, *Department of Engineering & Technology, Texas A & M University, United States*

Nakano, R., *Social Innovation Unit, Open Innovation Platform Headquarters, National Institute of Information and Communications Technology (NICT), Japan*

Otsuru, Go, *National Institute of Information and Communications Technology, Japan; Keio University, Japan*

Prasad, Ramjee, *CTIF Global Capsule, Herning, Denmark*

Radaydeh, Redha M., *Department of Engineering & Technology, Texas A & M University, United States*

Rohokale, Vandana, *Sinhgad Institute of Technology and Science, India*

S., Nandakumar, *School of Electronics Engineering, Vellore Institute of Technology, India*

Sanada, Yukitoshi, *Keio University, Japan*

Sen Rinku, *Department of Computer Science and Engineering, University of Calcutta, India*

Sharma, Swati, *Vishwaniketan Institute of Management, Entrepreneurship & Technology, India*

She, Xiaoming, *China Telecom Research Institute, China*

T., Velmurugan, *School of Electronics Engineering, Vellore Institute of Technology, India*

Tayem, Nizar, *Department of Engineering & Technology, Texas A & M University, United States*

Tsuji, Hiroyuki, *National Institute of Information and Communications Technology, Japan*

Tuna, Evren, *Department of Research and Development, ULAK Communications Inc., Turkey*

Wang, Jianxiu, *China Telecom Research Institute, China*

Yadava, Ramesh, *Department of Production and Industrial Engineering, BIT Sindri, India*

Yang, Shan, *China Telecom Research Institute, China*

Zhang, Xinxin, *School of Information and Communication Engineering, Beijing University of Posts and Telecommunications, China*

List of Abbreviations

3GPP	3rd generation partnership project
AI	Artificial Intelligence
AMPS	Advanced mobile phone systems
API	Application programming interface
AR	Augmented reality
ARoF	Analog radio over fiber
A-SCA	Area sweep coverage algorithm
ASCP	Area sweep coverage problem
ASR	Automated speech recognition
B5G	Beyond 5th generation
BBU	Base band unit
BH	Backhaul
BIRCH	Balanced Iterative Reducing and Clustering using Hierarchies
BN	Batch normalization
BPN	Basic psychological need
BS	Base station
CA	Conversational agent
CALL	Computer assisted language learning
CAPT	Computer-assisted pronunciation training
CBSM	Correlation-based stochastic models
CDF	Cumulative distribution functions
CDR	Call detail record
CEFR	Common European Framework of Reference
CFT	Clustering feature tree
CN	Core network
CNN	Convolutional neural network
CPRI	Common public radio interface
CPS	Cyber-physical systems
C-RAN	Centralized radio access network

CSI	Channel state information
CU	Central unit
D2D	Device-to-device
DBA	Dynamic bandwidth allocation
DC	Direct current
DCCS	Data-centric cloud service
DL	Deep learning
DL	DOWNLINK
DNN	deep neural network
DOA	Direction of arrival
DP	Deep processing
DP	Dynamic programming
DU	Distributed unit
EC	Edge computing
EE	Energy efficiency
EFL	English as a foreign language
eMBB	Enhanced mobile broadband
ER	Extended reality
ERP	Equally reduced power
ESL	English as a second language
ESPRIT	Estimation of signal parameters via rotational invariant techniques
ETSI	European Telecommunications Standards Institute
EV	Electronic vehicle
EVD	Eigenvalue decomposition
FH	Fronthaul
FLL	Foreign language learning
FPGA	Field-programmable gate array
GBM	Gradient boosting machine
GBSM	Geometry-based stochastic models
GCN	Graph convolutional network
GEO	Geostationary earth orbit
GNN	Graph neural network
GoB	Grid-of-beams
GRU	Gated recurrent unit
HOL	Head-of-line
HTS	High-throughput satellites
HW	Hardware

ICT	Information and communication technology
IoT	Internet of Things
IP	Interference pricing
JGN	Japan Gigabit Network
K-NN	K-nearest neighbor
L5G	Local 5G
LSTM	Long short-term memory
MBA	Multibeam antennas
MDP	Markov decision process
MF	Matched filtering
MIMO	Multiple input multiple output
ML	Machine learning
MMSE	Minimum mean square error
mMTC	Massive machine-type communication
MUSIC	Multiple signal classification
NFV	Network function virtualization
NI	National Instruments
NICT	National Institute of Information and Communications Technology
NLP	Natural language processing
NN	Neural network
NOC	Network operations center
NR	New radio
NTN	Nonterrestrial network
ODN	Optical distribution network
OFDM	Orthogonal frequency-division multiplexing
OFDMA	Orthogonal frequency division multiple access
OLT	Optical line terminal
ONF	Open Network Foundation
ONU	Optical network unit
OS-ELM	Online sequential extreme learning machine
OWC	Optical wireless communication
PCA	Principal component analysis
PCI	Physical cell ID
PLS	Physical layer split
PMI	Precoding matrix indicator
PPS	Pulse per second
P-SCA	Point sweep coverage algorithm
QAOA	Quantum approximate optimization algorithm

QFS	Quantum feature selection
QML	Quantum machine learning
QNN	Quantum neural network
QoE	Quality of experience
QoL	Quality of life
QoS	Quality of service
QPCA	Quantum principal component analysis
QSVD	Quantum singular value decomposition
QSVM	Quantum support vector machine
RAN	Radio access networks
RB	Resource block
RDEESA	Real time dynamic efficient spectrum analysis
RE	Resource elements
RF	Radio frequency
RRH	Remote radio head
RS	Reference signal
RSRP	Reference signal received power
RSRQ	Reference signal received quality
RTT	Round-trip time
SDN	Software-defined networking
SDR	Software-defined radio
SDS	Spoken dialogue systems
SE	Spectral efficiency
SINR	Signal-to-interference-plus-noise ratio
SLA	Service level agreement
SMS	Short message service
SP	Strict priority
SSB	Synchronization signal block
SSS	Secondary synchronization signals
SVD	Singular value decomposition
SVM	Support vector machine
THz	Terahertz
TSP	Traveling salesman problem
UAV	Unmanned aerial vehicle
UE	User equipment
ULA	Uniform linear array
URLLC	Ultrareliable and low latency communication
VCS	Video conference system

VQE	Variational quantum eigensolver
VR	Virtual reality
ZF	Zero forcing

Part I

Technologies

DOI: 10.1201/9788743801108-1

1

Opportunities and Limitations of Dialogue-based Foreign Language Learning AI Applications

Despoina N. Gkountara[1], Ramjee Prasad[2], and Pavlos I. Lazaridis[1]

[1]University of Huddersfield, United Kingdom
[2]CTIF Global Capsule, Herning, Denmark
E-mail: d.gkountara@hud.ac.uk; p.lazaridis@hud.ac.uk;
prasad.ramjee@ctifglobalcapsule.org

Abstract

The utilization of artificial intelligence (AI) is an undeniable fact and is quickly revolutionizing the field of education, through its many capabilities, by changing the scenery of when and how people teach and learn. Especially with the emergence of the 5G and 6G (fifth/sixth-generation wireless communications) and their substantially higher speeds and much lower latency, as well as new faster and more powerful smartphones capable of running advanced applications, AI-powered educational tools are becoming even more sophisticated and efficient. This has a clear application in the spectrum of foreign language learning (FLL). 5G/6G enhances the opportunity for a ubiquitous and immersive learning experience, for example, through smart and immersive classrooms, distance one-to-one learning, numerous learning smartphone applications, and many others. One of the many facets of AI in FLL is dialogue-based language learning, in the form of intelligent computer-assisted language learning systems, conversational agents, pronunciation training tools, and chatbots. All these systems require increased wireless bandwidth since they require virtual presences, robots, and high-resolution features. Dialogue-based language learning tools are becoming predominant and gradually more people are using them. While AI in education is facing

DOI: 10.1201/9788743801108-2

unprecedented development, much improvement is still necessary, in aspects such as performance, user-friendliness, pedagogical effectiveness, and ethics. This chapter investigates the basic functions of language learning applications and chatbots that are powered by AI. Recent advances made in this area and some suggestions for future development are presented.

Keywords: Adaptive learning, artificial intelligence, chatbots for foreign language learning, conversational agents, fifth-sixth generation (5G/6G) networks, intelligent computer-assisted language learning, machine learning

1.1 Introduction

In our era of globally spread information and constant digital interconnection, the process of acquiring new knowledge has already been taking a new form in everyday human activity. Computer systems and their possibilities are modifying all aspects of our lives, and teaching and learning cannot be left uninfluenced by the opportunities they provide to transform the learning process. 5G/6G technologies provide many possibilities for language immersion, such as high-speed wireless networks in classrooms, virtual reality interactions, and endless practice opportunities. As 5G moves on to be broadly adopted and implemented, and 6G is under development, AI applications for language learning will be more and more efficient. Applications such as haptic communication with massive ultrareliable-and-low-latency-communications and holographic communications [1] will enhance reliability and collaborative cognitive communication systems (the 3C systems) [2] are gradually becoming more and more possible. In this context, the educational process will be shifting into a more human-centric approach to teaching, as learning can be transferred from a classroom to anywhere the learner has access to an internet connection and ubiquitous learning is gradually transforming both the teacher and the learner.

From the point of view of foreign language learning, the latest technological possibilities are particularly important, because their promised affordances can increase the practice opportunities in a more realistic environment. With the evolution of speech recognition and natural language processing, learning a second language is shifting to a student-centered pedagogical model, as modern pedagogy proposes. Various computer applications supporting task-based learning and collaborative practicing are developing into a commonplace, and this seems to be the future trend in foreign language learning (FLL). The latest developments around deep learning in the last decade are further boosting this tendency, which comes with new promises, but also with the need for the necessary adjustment.

The first chatbots were not designed for educational purposes. For example, ELIZA the chatbot (the first chatbot, designed in 1966) was a tool for the emotional support of its users. Two of the early chatbots that were used for language learning were Cleverbot (launched in 2008 by Rollo Carpenter) and Elbot (created by Fred Roberts and Artificial Solutions in 2001) which were designed as conversational assistants that are constantly improving by using machine learning. Microsoft's chatbot called XiaoIce (an "empathetic" chatbot, designed in 2014) is used in the United States, Japan, China, and Indonesia. It utilizes neural networks and deep learning, has millions of users, and has become more efficient in interacting with humans. It is a poet, a singer, a children's audiobook reciter, a show host, and generally, an "AI being" [3].

By analyzing chatbots for language learning, it becomes obvious that despite advances in natural language processing (NLP) and deep processing (DP), each considered crucial in advancing AI technology, it is unclear if chatbots have reached their peak in communication skills, especially for advanced student levels. Chatbots base their interactive functions on the millions of interactions and dialogues that have taken place previously and constantly learn from them. Although some researchers claim that chatbots are not yet able to efficiently teach foreign languages to an advanced level, research indicates that the majority of chatbots can improve the learner's reading competence, narrative skills, comprehension, and pronunciation, and they can reduce anxiety levels, their social presence, and collaboration. According to Dokukina and Gumanova [3], the effectiveness of chatbots derives from: (1) the effect of "microlearning," that is, the division of compound subjects of knowledge into smaller parts, easier to grasp and with the possibility of being visited anytime; (2) the combination of visual and audio strategies; (3) the fun-way of learning, that is, gamificated learning enriched with task-based learning activities; (4) the motivating tools, such as automated reward systems, or practice reminders; and finally, (5) the capability to learn at all places and times (ubiquitous learning).

With the increasing use of "AI beings," people appear interested in learning languages via avatars and computers, or a hybrid combination of humans and computers. Especially in the last decade, the improvement of technologies such as automated speech recognition (ASR) and text-to-speech conversing systems have assisted chatbots to become more independent and competent communicators, as well as language learning partners. A very eloquent example of this fact is the case of Kuki, a part of a conversational AI agent, previously known as Mitsuku (created and maintained by Steve Worswick in 2004). Kuki is presented as an 18-year-old female persona, who has become an "influencer," after having inspired movie directors (the movie HER was inspired by Kuki), and even as a virtual model for the clothing industry. Kuki has been five times awarded the Loebner price for successfully passing the Turing Test, for a computer being capable of human-like thinking. The above tools have facilitated the users' lives and are being used by hundreds of millions of people daily [4].

Researchers appear to agree that chatbots can be a useful tool for foreign language learning (FLL) but also one that is in its first stages of study and can be improved. Some of the chatbots' limitations are that, at this moment, they cannot yet generate answers to all levels of unstructured inputs from the end-users. For the time being, however, many of the chatbots are useful for early to intermediate language learning stages. For example, a Russian English-learning chatbot, Vasya, is an example of a mobile application designed to teach English via text, audio, and video instructions. Vasya has a good sense of humor, but it is also an efficient tool for simple explanations to beginner-level English learners [3]. Despite all the advantages presented above, the numerous manifestations of AI in education are placing the learning process in front of new challenges.

1.2 Artificial Intelligence Implementations in Foreign Language Learning

AI implementations in education cover all aspects of the spectrum of language education, from the student's, as well as the teacher's perspective, but also systems used in schools and universities, for student filing, progress prediction, and customized advice and support. Baker and Smith [5] attempt a categorization that divides AI tools used in education into three groups: (1) learner-facing, (2) teacher-facing, and (3) system-facing systems. Learner-facing AI systems come in the form of software utilized by students to acquire knowledge. Teacher-facing systems are utilized by tutors to reduce their workload and make their work more effective in domains such as administration, assessment, feedback, and plagiarism detection. System-facing AI tools provide information for administrators and managers on the institutional level, for example, they help monitor attrition patterns across faculties or colleges.

In the domain of foreign language learning, Pokrivčáková [6] adopts a basic division of AI implementations in foreign language learning: (1) the creation of customized, adaptive educational packages that are in a position to follow the learners' requirements and progress; (2) the use of machine translation mechanisms; (3) the design of writing assistants to support writing skills; (4) the utilization of chatbots, for the improvement of listening and speaking skills; (5) special AI-powered online language learning applications and platforms; (6) the intelligent tutoring systems, as an evolution of the traditional one-to-one teaching, which take decisions after elaborating relevant data concerning student performance and progress; (7) the development of specially designed systems for cooperative learning; and (8) the implementation of intelligent virtual reality systems for creating an authentic learning context. The above-presented, multifaceted realization of AI in language learning processes outlines an affluence of potential that progressively will be incorporated in all aspects of language learning.

The dialogue-based conversational AI-powered agents are based on the natural interaction between the learner and a virtual agent controlled by the computer through artificial intelligence. This aspect of AI has been proven to have important advantages on conversational skills, pronunciation accuracy, control of discomfort, and customization to the needs and progress of the learner. An example of incorporating conversational AI in the form of a "robot in the classroom" named Interact4School was a specially designed program by Torben Schmidt and Thomas Strasser [7]. This program has been tested in schools in Germany, initiating an automated system named FeedBook. High school students worked with FeedBook alongside their usual program of language learning. The experiment provided evidence that learners had better results after receiving immediate, specific feedback from the automated system. The authors consider as the main pillars of artificial intelligence in foreign language learning: a domain model (the grammatical and syntactical rules of the target language), a constant error analysis system (machine learning for continuous update of student errors), and a learner model (data concerning student progress and preferences). The goal of the program was to keep the learners within their "zone of proximal development," defined as the space between what the learners can do with the instructor's oversight and what without any assistance. This way, the learner is constantly progressing from needing-to-not-needing-assistance when doing language tasks. Finally, Schmidt and Strasser support the idea that AI systems are best utilized when they give a communicative aspect to the tasks so that the forms of the target language are learned implicitly, informally, and with a communicative purpose.

A major systematic review of research on the field of dialogue-based FLL by Bibauw et al. [8] aimed to define the boundaries and the major research results in this area. After analyzing 207 peer-reviewed studies, spanning from 2007 to 2017, they investigated how the learner interacts with a computer in the second language (L2), with a focus on learning. The criteria of study selection were the following: (1) the learner is expected to be actively involved in a dialogue only with a virtual agent, instead of a human instructor; (2) the language used is the target language, instead of the native language; and (3) the dialogue is the main task, as opposed to item-based learning of many AI language learning applications, which are focusing on specific language phenomena and grammatical forms. They concluded that there are four major strands: (1) the so-called "Intelligent CALL" (ICALL), defined as "systems to provide software that interacts intelligently with students, responding flexibly and dynamically to student's learning progress" [9]; (2) the computer-assisted pronunciation training (CAPT) systems [10], defined as systems designed to help students practice speaking foreign languages by offering automatic pronunciation assessment and corrective feedback; (3) the spoken dialogue systems (composed by a speech recognizer in combination with a text-to-speech module) and conversational agents (SDS/CA), which are dialogue systems employing one or more of: text, speech, graphics, haptics, gestures, and other modes of communication on both the input and output channel [11]; and (4) the chatbots, which are systems capable of maintaining a conversation with a user in natural language and simulating the way a human would behave as a conversational partner [12].

Huang et al. [13] support the idea that chatbots can be optimized if we point out exactly the areas in which their performance can be improved. In a recent systematic review of 25 empirical studies [13], the authors investigated the technological, pedagogical, and social advantages of chatbots, from the aspect of foreign language learning. They searched if chatbot "usability" can be enhanced by their "utility." One very interesting fact is that most interest is manifested from Asian researchers, which indicates how important is for these countries to learn the English language in a faster and more effective way. The results of the study included the timeliness, ease of use, and personalization of the learning experience. Most pedagogical advantages derive from the chatbots' function as: (1) conversational partners (functioning as practice interlocutors); (2) as simulators, for example in the form of an English-speaking persona, to create an authentic speaking context (e.g., in a restaurant or a grocery store); or (3) as helplines. They claim that chatbots will have many more users if their technology is more user-friendly and technologically advanced [13].

Haristiani [14] attempts a categorization of chatbots depending on their structure, purpose, and target audience. Structurally, she identifies as tree-based, "Flow" chatbots those whose responses come from a fixed information pool (database); these chatbots cannot participate in any dialogue that does not refer to this material. A second category of "artificially intelligent" chatbots is in a position to reform their understanding by utilizing information from former discussions and users' input. This kind of potential corresponds to a more liberated form of interaction with end-users. Finally, Haristiani describes "hybrid" as the ones that merge the flow and artificial intelligence mechanisms in one chatbot, which has the potential to expand their knowledge and interactive possibilities, but at the same time to stay close to the creator design in terms of communication. The main and more interesting characteristic of conversational AI chatbots is that they do not interact with end-users in a predetermined way. Chatbots usually select their responses heuristically through discovery, and by using fuzzy logic (an approach to processing that allows a variable to assume several possible truth values). As to their purpose and utility, Haristiani [14] divides chatbots into "entertaining" and "functional" chatbots, while, in relation to the audience, she characterizes as "generalist" those who mainly provide general information and assistance, while the "specialist" chatbots offer extremely specialized assistance, such as ordering online. Educational chatbots, as investigated by Haristiani, mostly belong to the AI "family" of chatbots. She indicatively presents the MILA bot, a chatbot that can converse naturally with users, explaining that this kind of conversation can happen when that chatbot incorporates features such as template-based models, bag-of-words models, sequence-to-sequence neural network, and latent variable neural network models [14].

In another study, Belda Medina and Calvo-Ferrer investigated the impact of conversational AI among 176 student teachers during a four-week time span in Spain (N = 115) and Poland (N = 61). The study identified a generally positive attitude toward AI tools. More specifically, the supposed ease of use and confidence were higher, whereas the strength of intention to use AI conversational tools was lower. Also, a higher satisfaction was demonstrated among female teacher-trainees in chatbots' areas of interest, compared to male trainees. Moreover, a lower contentment was identified in male trainees compared to their female cotrainees [15].

A conversational agent is a broad concept that includes chatbots, but also virtual assistants (such as Alexa and Siri). The dialogue-based tools have slight differences, and often they overlap in functions. For example, spoken dialogue systems (SDS) and conversational agents (CA) utilize similar technologies. Jeon et al. [16] state that the broad spectrum of diversified chatbots is currently leading to a not very clear categorization, therefore, it becomes difficult for researchers to proceed to safe conclusions about efficiency, but also it becomes more difficult for teachers and learners to choose the best chatbot for their learning and teaching needs. Jeon states that until today, two main areas have been uncovered by research in the domain of language learning speech-recognition chatbots: (1) the differences between speech-based and text-based chatbots: so far, the two types of chatbots are analyzed as one category; and (2) possible variations deriving from the special design characteristics of language learning chatbots. These two aspects need to be considered when striving to understand the educational effects of one chatbot on another, as Jeon supposes that different chatbots create different educational experiences. From the chatbot design point of view, after the examination against analytical protocols, Jeon et al. focused exclusively on empirical studies that concerned speech-recognition chatbots in the last 20 years. The authors reflected on three criteria: (1) the goal orientation (i.e., if there are specific educational aims for the creation of the chatbot), (2) the embodiment (i.e., some form of visual illustration), and (3) the multimodality (i.e., if there are more than one different mechanisms for conversation with the user). They named the criteria "GEM" criteria, from the initials goal, embodiment, and multimodality.

1.3 AI Chatbots for FLL

The new smart conversational AI agents are considered a new array of evolved chatbots, which can help to forecast customer needs, based on their past conversations, and purchase or browsing behaviors [17]. Chatbots are a set of systems and applications, with specific intelligence, scope, emotional capacity, and flow of conversations, and they support the idea of utilizing a computer to converse with a human interlocutor, with a specific purpose. Some of the most popular language-learning applications today are: Duolingo, Memrise, Mondly, Lingvist, Babbel [18], Andy, Eggbun, Rosetta Stone, Busuu, and HelloTalk, ranking in the top places with regard to the number of active users. Some other ones are: Tandem, Super Duolingo, Drops, Lirica, Pimsleur, LingusLift, Mindsnacks, TripLingo, MosaLinga,

HiNative, (How to) Pronounce, Speakly, and QLango, just to name a few. The most recent multifunction chatbot is ChatGPT.

Indicatively, one of the most popular applications – if not the most popular – *Duolingo*, was first launched in November 2011 but formally began to teach Latin-American Spanish in 2013. It was written in the Python programming language. Duolingo currently offers 43 different languages on all mobile platforms, 39 of which are for English speakers. Some of the languages taught are better supported than others, and this correlates with the demand for the language. This fact is due to the nature of the AI function, which becomes richer as the interest (and the consequent input) from learners increases, as the AI component can be improved by the users' input. Although it focuses more on writing and less on speaking practice, Duolingo's roleplay feature (powered by Duolingo Max) opens an AI chatbot that permits its learners to exercise their real-world discussion capabilities with characters in the application. Pedagogically, it is based on a reward and motivation system that constantly follows students' progress, to propose appropriate material. Duolingo's entertaining appearance has catapulted it to the top of learners' preferences [19, 20]. A wealth of studies starting from 2012 reveals Duolingo's impact on users' motivation, but also learning outcomes. Just to name one of the empirical studies, a recent inquiry on university students in Indonesia showed positive results on the students' scores after a 30-day use of Duolingo. This is only one of the numerous studies that have taken place on Duolingo application in the period after its creation.

Babbel (launched in January 2008) gained millions of subscribers very fast. This is a clear indicator of people's interest in saving time and money with the aid of technology by investing their time in learning languages online. This tendency has only been accentuated with the advent of COVID-19. Babbel's users can learn up to 15 languages, through over 60,000 lessons, generated by 200 specialists in relevant pedagogy. The learners' performance is constantly examined by the application components, with the aim to reformat the trainee participation. The Babbel application currently grants its users the potential to rehearse new words during their time spent online on the internet, by reflexively interpreting specific terms and expressions on the page visited into the target language. This concept allows the learner to absorb new vocabulary without the obligation to assign a specific time "slot" for language learning, while at the same time, they can practice more the words that they did not have the time to learn previously [21, 22].

An example of a very popular application that is mainly based on AI is the *Mondly* chatbot, launched in 2014. Mondly is characteristic because of the possibilities it offers; it teaches 41 languages and has already been downloaded by tens of millions of learners in 190 countries. Mondly was designed to interact with users, understand voice input, and reply with a human voice, while being engaging, but also entertaining. Apart from speech recognition technology, it also implements new technologies, such as virtual reality (VR) and augmented reality (AR). It also uses virtual characters and simulations of real-life situations to practice speaking. The Mondly's "brain" is a key-phrase database. In order for Mondly to appear smarter and be a more independent interlocutor, its new upgrades focus on the use of machine learning systems. Several studies see some weaknesses in Mondly, suggesting that it is efficient for beginner students only and propose complementary tools as compensation [23, 24].

Artificial intelligence tools make good use of proven learning theories, such as the "spaced repetition" learning model. An example of this fact is *Memrise*, which is a British language-learning application that uses flashcards. These artifacts are presented according to the spaced repetition theory to increase efficiency. This technique is blended with an "AI language partner," activated by a generative pretrained transformer that permits users to practice human-like conversations. The advantage of Memrise stands in the enhancement of self-assurance of the user in the speaking domain. Numerous research projects explored the effect of the use of Memrise on vocabulary retention. Indicatively, a study on 75 intermediate-level English students, against a control group of students who were taught English vocabulary in the traditional way, showed better results in immediate post-test scores, and collocation in the English language [25]. A study on Vietnamese students showed similar results on vocabulary [26]. Memrise was tested as an additional learning subject on a French language learning group of A2 level university students who wished to improve their listening skills in the French language. Pre-test, post-test, survey, and observations, provide an indication that the use of Memrise as a complementary tool improved understanding of French audio texts [27]. Numerous more studies have concluded that Memrise is a good option for supplementary use and that it can yield good results in vocabulary skills, from an early age [28] to adult university language students [29].

In November 2022 the new multifunction chatbot *ChatGPT* was launched by OpenAI, and by January 2023, it had gathered an impressive number of 100 million users. Without being specifically designed to be a language learning application, ChatGPT's capabilities extend to also being a foreign language conversational partner. According to the study of Kohnke et al. [30],

ChatGPT can explain unknown words, correct mistakes, create texts, develop quizzes, and generate dialogues by adjusting the difficulty level to the different levels of language knowledge [31]. All these attributes provide teachers with valuable tools, for example, in assigning differentiated tasks. Fears that ChatGPT can facilitate plagiarism and cheating have been recently tackled by a new update of Turnitin, the most effective version of which can detect AI-generated academic texts [32]. Other reservations refer to possibly inaccurate information provided by ChatGPT – for example, in presenting syntax uses of words. This kind of mistake is due to the mere nature of the AI mechanism in searching for information, which is, as mentioned, based on the plausible and the possible, and not the most reasonable answer [33].

Very characteristic in structure is an experimental chatbot named *AsasaraBot*, developed by Greek researchers and tested in two private language schools in Greece with the goal to familiarize students with both Greek mythology aspects and two foreign languages, English and French [34]. Its function is based on a natural language processing (NLP) part, a dialogue managing component, and an information retrieval and knowledge base element, each module playing a role in the communication with end-users. 61 participants were invited to evaluate the program by replying to pretest and post-test questionnaires. The aim of the project identifies very specifically the challenges in the use of chatbots and aims to continue the investigation of the right balance between ICT-powered learning and conventional, traditional teaching.

5G networks have been used in virtual reality education, to form comprehensive classrooms, and to provide immersive experiences to homeschooled students or students with special needs. As an example of the implementation of 6G network capabilities, it is worth referring to an intuitive phrase recognition system that was developed using machine learning, to achieve an organic dialogue to teach Mandarin language. Combined with virtual reality (VR) techniques, it is a tool generated for Mandarin language learning. The outcomes exhibit high levels of engagement and the efficacy in learning was better than in traditional ways of teaching such a challenging language [35].

1.4 Perceptions, Challenges, and Suggestions Regarding the Use of Chatbots as Language Learning Tools

Some researchers insist that chatbots work their best when combined with other learning tools and techniques, such as gestural data. There is a researchers' shared view that for the time being, the combination of "traditional" and "novel" is the best way to achieve the best learning results. This view is proposed by Ji et al. [36], who analyzed systematically 24 empirical studies that describe the human–chatbot collaboration during the utilization of chatbots in language education. The review found that in a large number of studies, chatbots were not utilized as the only learning tool, without teachers' involvement. Ji et al. recommend the complementary utilization of chatbots in order to control the chatbot technology-related hindrances that may occur. This review on conversational agents investigated the teachers' views on how to incorporate conversational agents into their English as a second language (ESL) program. They found that generally, the input was positive, although some improvement on the interface and user-friendliness was required.

Several studies indicate that language students welcome chatbots as a promising tool for language learning. This is indicated by most studies that have been conducted all through the last decade. As advantages, most studies present the following: (1) endless practice opportunities, (2) anxiety control, (3) improved levels of speaking confidence, (4) immediate and effective feedback, (5) provision of a more authentic speaking context, (6) exposure to a broader variety of language structures and vocabulary, (7) entertaining way of learning, and (8) ubiquitous learning. As drawbacks, many studies refer to (1) the "novelty effect," which refers to the user's gradual loss of interest as the user gets more accustomed to the technological features; (2) inaccurate information provided; and (3) low levels of user-friendliness. Chatbots are yet in a stage of improvement: a study on Colombian university students using Memrise as a vocabulary retention tool showed ambiguous results on the level of interest, despite the language learning results being improved. This result reveals a need for improvement in user-friendliness [37–40].

Shabani [41] conducted a descriptive content analysis to gather intermediate-level language-learning subjects of the five most popular language-learning CALL applications: Duolingo, Babbel, Busuu, Memrise, and HelloTalk. He perceives three basic psychological needs (BPNs) as the most important factors during the use of language learning applications: (1) autonomy, (2) competence, and (3) relatedness – and also, combinations thereof. Autonomy refers to the language learning application considering the learner's progress and letting them independently select tasks and assignments. Competence is about creating appropriate task content, offering feedback to learners, and generally motivating learners by exposing their positive results. Relatedness denotes how the language learning application relates to the learner's standards. Results showed that all the five applications investigated addressed the three BPNs and their combinations, with Duolingo and Babbel being more competent in all aspects.

In Murray's survey [32], conducted by teachers with 18 student participants, the educational experience of ChatGPT has been described as "not great" but also "neither bad." The most interesting aspect of their opinions was that young students welcome AI technology, by accepting that this tool is here to stay and that they wish to know it as best as they can so that they will learn how to use it in an unbiased manner, without "mythologizing" it, because preconceptions can only lead to misuse.

In another study [42], six popular virtual reality applications for foreign language learning are analyzed in higher education contexts: Mondly, VRSpeech, VR Learn English, Gold Lotus, AltSpaceVR, and VirtualSpeech. It is stated that the preferred VR application for foreign language learning integrates elaborate virtual context for full concentration, high level of optical impressions, straightforward control of the virtual personas, rigorous subject matter choice, high level of relevance and collaboration in all cases, cost-effectiveness, and lastly, discreet formative assessment of user learning.

A recent review of 15 empirical studies of a total of 668 university EFL (English as a foreign language) programs, in seven different learning environments in Asian and European countries and Egypt, investigated the pedagogical gains from the use of chatbots against widely accepted pedagogical criteria [43]. The pedagogical correlations indicated that chatbots have been found to be able to assess the user's level of language proficiency within the Common European Framework of Reference (CEFR). Furthermore, chatbots encourage self-regulated learning by enhancing the individual learner's experience, by enhancing cognitive, motivational, behavioral, and affective variables. As for the situated learning theory, where the acquisition of knowledge happens within a collaborative informal context in which learning occurs unconsciously, there was a positive correlation between chatbots and learning.

From the aforementioned studies, it becomes clear that chatbots in general give positive results in the frequency of students speaking, practicing, and engaging in learning. This does not mean that AI-empowered educational tools do not need more investigation, both in research and in practice. Their encouraging effects on the students' opportunities for interaction in a new language, the informality of the learning context, but also the levels of the students' influence on their personal learning experience require a more thorough investigation. These are aspects that are too important to ignore, especially in our era of educational reforms [44].

Two of AI's mechanisms are supervised and reinforcement learning. Supervised training is based on specifically labeled data sets, while reinforcement learning is a capacity to associate with the context and function independently in the search for new knowledge. Thus, conversational agents choose the most probable answer from the available data. This means that the more input they receive, the more probable it becomes for their mechanism to come up with the correct answer [45, 46]. One recent teacher's concern is that AI technology can lead to "fake" results that cannot be easily distinguished by the detecting applications that are currently available. There is a necessity to become familiar with AI's mechanisms to better understand AI and tackle some possible downsides it presents. It becomes necessary for teachers and learners to understand that AI-powered chatbots, in general, do not utilize logical sense, but possibilities.

1.5 Conclusion

5G/6G wireless networks are providing a new environment for experimentation on what artificial intelligence can provide for education. However, more research is needed before we proceed to safe conclusions as to how teachers can find ways to adopt AI technology into their practice. The emergence of AI as a foreign language tool, along with its high popularity, especially among young people, requires teachers to have a thorough and deep understanding of the opportunities it provides. As network and smartphone capabilities increase with 6G networks, more reliable tools will be crafted. Furthermore, it is understood that a necessary shift in strategy is due, which requires relevant and constant training. In this short review of the latest implementations of chatbots in language learning, there are numerous studies suggesting that chatbots and conversational agents for language learning attract many users and perform well with beginner and intermediate-level students.

With the surge in chatbot research and development, there is still much confusion as to the criteria that chatbots must meet to offer educational results in language learning, as well as to which is the best way to incorporate AI into educational settings. It is important to deeply understand what AI chatbots are and what they are not. Surely, chatbots do not rely on reason, and this fact advocates the general view that a hybrid combination of human instruction and an educated utilization of chatbots is still the best way to enhance engagement and improve results in students of all levels. In this stage, it clearly emerges that chatbots have a promising future in FLL, but also that serious and thorough research is needed in several areas, such as user-friendliness, accuracy, and ways of best use. Moreover, there is an urgent need for appropriate teacher training to eliminate any negative mythology on AI potentials and dangers, and for instructors to be able to tackle emerging issues during the use of AI in teaching and learning.

Acknowledgment

This research was partially funded by the European Union, through the Horizon 2020 Marie Skłodowska-Curie RISE Programme "Research Collaboration and Mobility for Beyond 5G Future Wireless Networks (RECOMBINE)" under grant agreement no. 872857, and through the National Recovery and Resilience Plan of the Republic of Bulgaria, BG-RRP-2.005 – "Twinning" with Project No. BG-RRP-2.005-002 titled "Twinning for Excellence in Research in Sustainable Future Communication Networks in the Context of a Green Economy – GREENBEAT."

References

[1] Y. Sharrab, N. T. Almutiri, M. Tarawneh, F. Alzyoud, A. R. Al-Ghuwairi, & D. Al-Fraihat, "Toward Smart and Immersive Classroom based on AI, VR, and 6G". Int. J. Emerg. Technol. Learn., vol. 18(2), pp. 4-16, 2023.

[2] S.Shahzadi, M. Iqbal, N. R. Chaudhry "6G vision: Toward future collaborative cognitive communication (3C) systems", IEEE Communications Standards Magazine, vol. 5(2), pp. 60-67, 2021.

[3] I. Dokukina, and J. Gumanova, "The rise of chatbots–new personal assistants in foreign language learning", Procedia Computer Science, vol. 169, pp. 542-546, 2020.

[4] S. Pasang, A. Latifa, & S. Ammade, "The Use of Kuki Chatbot Application to Improve English Achievement", International Journal of English Language Studies, vol. 6(1), pp. 15-46, 2024.

[5] T. Baker, L.Smith, N.& Anissa (2019). "Educ-AI-tion rebooted? Exploring the future of artificial intelligence in schools and colleges". Retrieved May, 12, 2020.

[6] S. Pokrivčáková, "Preparing teachers for the application of AI-powered technologies in foreign language education", Journal of Language and Cultural Education, vol. 7, pp.135-153, 2019.

[7] T. Schmidt, T. Strasser, "Artificial intelligence in foreign language learning and teaching: a CALL for intelligent practice", Anglistik: International Journal of English Studies, vol. 33.1, pp.165-184, 2022.

[8] S.Bibauw, T. François, P.Desmet, "Discussing with a computer to practice a foreign language: Research synthesis and conceptual framework of dialogue-based CALL" Computer Assisted Language Learning, vol. 32, pp. 827-877, 2019.

[9] J. Gamper, J. Knapp, "A review of intelligent CALL systems", Computer Assisted Language Learning, vol. 15(4), pp. 329-342, 2002.

[10] P. M. Rogerson-Revell, "Computer-assisted pronunciation training (CAPT): Current issues and future directions". Relc Journal, vol. 52(1), pp. 189-205, 2021.

[11] C. Troussas, A. Krouska, M. Virvou, "Integrating an adjusted conversational agent into a mobile-assisted language learning application", In 2017 IEEE 29th International Conference on Tools with Artificial Intelligence (ICTAI), pp. 1153-1157, Nov. 2017.

[12] L. Fryer, D. Coniam, R. Carpenter, D. Lăpuşneanu, "Bots for language learning now: Current and future directions", Language Learning &Technology, vol. 24(2), pp. 8-22, June 2020.

[13] W. Huang, K. F. Hew, L. K. Fryer, "Chatbots for language learning—Are they really useful? A systematic review of chatbot-supported language learning", Journal of Computer Assisted Learning, vol. 38(1), pp. 237-257, 2022.

[14] N. Haristiani, "Artificial Intelligence (AI) chatbot as language learning medium: An inquiry", Journal of Physics: Conference Series, vol. 1387, No. 1, p. 012020, IOP Publishing, November 2019.

[15] J. Belda-Medina, and J. R. Calvo-Ferrer, (2022), "Using chatbots as AI conversational partners in language learning", Applied Sciences, vol. 12(17), 8427, August 2022.

[16] J. Jeon, S. Lee, and H. Choe, "Beyond ChatGPT: A conceptual framework and systematic review of speech-recognition chatbots for language learning", Computers & Education, 2023.

[17] L. Vishnoi, "Conversational agents: a more assertive form of chatbots", published in Towards Data Science, Jun. 2020.

[18] S. Loewen, D. R. Isbell, and Z. Sporn, "The effectiveness of app - based language instruction for developing receptive linguistic knowledge and oral communicative ability", Foreign Language Annals, vol. 53(2), pp. 209-233, 2020.

[19] J. Petrović, and M. Jovanović, "The role of chatbots in foreign language learning: the present situation and the future outlook", Artificial Intelligence: Theory and Applications, pp. 313-330, 2021.

[20] M. Shortt, S. Tilak, I. Kuznetcova, B. Martens, & B. Akinkuolie, "Gamification in mobile-assisted language learning: A systematic review of Duolingo literature from public release of 2012 to early 2020", Computer Assisted Language Learning, vol. 36(3), pp. 517-554, 2023.

[21] Z. Sporn, J. Chanter, & D. Meehan, "Babbel Language Learning Podcasts", International Journal of Advanced Corporate Learning, vol. 13(3), 2020.

[22] M. Mendes de Oliveira, Z. Sporn, L. Kliemann, A. Borschke, & M. Meyering, "Online language learning and workplace communication: a study on Babbel's virtual-classroom solution", Computer Assisted Language Learning, vol. 36(8), pp. 1501-1527, 2023.

[23] M. Nushi, N. Fattahi and F. Ebn-Abbasi, "An Evaluative Review of Mondly: A Mobile Language Learning Application", The EuroCALL Review, vol. 30(2), pp. 69-85, 2023.

[24] S. Hajizadeh, A. R. Salman, & S. Ebadi, "Evaluating language learning applications from EFL learners' perspectives: The case of Mondly", preprint on Researchsquare, 2023.

[25] Z. Esmaeili, & M. Shahrokhi, "The Impact of Memrise Application on Iranian EFL Learners' Collocation Learning and Retention", International Journal of Language Education, vol. 4(2), pp. 221-233, 2020.

[26] P. T. T. Nhu, and T. N. Q. Quyen, "Improving vocabulary learning with Memrise mobile application for students in Vietnam", International Journal on E-Learning Practices (IJELP), vol. 4, pp. 61-75, 2021.

[27] N. R. H. P. Paradhina, and M. Laksman-Huntley, "The Efficacy of Memrise as a Supplementary Study Material in Understanding French as a Foreign Language", In International Seminar on Language, Education, and Culture (ISoLEC 2021), Atlantis Press, pp. 249-252, December 2021.

[28] Nguyen, A. T., Nguyen, T. T., Le, T. T., Phuong, H. Y., Pham, T. T., Huynh, T. A. T., & Nguyen, H. T., "Effects of Memrise on Vietnamese EFL Students' Vocabulary: A Case Study at a College in a Rural Area", Electronic Journal of e-Learning, 21(5), 450-460, 2023.

[29] H. G. Valencia, G. D. I. Gómez, M. A. Idarraga, and M. R. Villaquiran, "Integrating the virtual platform Memrise as a teaching tool for English vocabulary in foreign language students of second semester

at a university", Latinoamericana de Estudios Educativos, vol. 16(1), pp. 259-284, 2020.

[30] L Kohnke, B. L. Moorhouse. and D. Zou, "ChatGPT for language teaching and learning", RELC Journal, vol. 54(2), pp. 537-550, 2023.

[31] W. C. H. Hong, "The impact of ChatGPT on foreign language teaching and learning: opportunities in education and research", Journal of Educational Technology and Innovation, vol. 5(1), 2023.

[32] Nancy Murray, "ChatGPT in the English Classroom: 18 Students Test it and Share their Insights." Teaching and Learning Excellence through Scholarship vol. 3(1), 2023.

[33] D. R. Cotton, P. A. Cotton, and J. R. Shipway, "Chatting and cheating: Ensuring academic integrity in the era of ChatGPT", Innovations in Education and Teaching International, pp. 1-12, 2023.

[34] K. Mageira, D. Pittou, A. Papasalouros, K. Kotis, P. Zangogianni, and A. Daradoumis, "Educational AI chatbots for content and language integrated learning", Applied Sciences, vol.12(7), p. 3239, Mar. 2022.

[35] M. Nomoto, A. Lustig, R. Cossovich, and J. Hargis, "Qilin, a robot-assisted Chinese language learning bilingual chatbot", In Proceedings of the 4th international conference on modern educational technology, pp. 13-19, May 2022.

[36] H. Ji, I. Han, Y. Ko, "A systematic review of conversational AI in language education: Focusing on the collaboration with human teachers", Journal of research on technology in education, Advance online publication, pp. 48-63, 2023.

[37] J. M. Tommerdahl, C. S. Dragonflame, and A. A.Olsen, "A systematic review examining the efficacy of commercially available foreign language learning mobile apps", Computer Assisted Language Learning, vol. 1-30, 2022.

[38] P. Ajisoko, "The use of Duolingo apps to improve English vocabulary learning". International Journal of Emerging Technologies in Learning (iJET), vol. 15(7), pp.149-155, 2020.

[39] Y. Tuerel, K., and E. Davudova, "Evaluation of mobile applications in foreign language learning at early age", Journal of Educational Technology and Online Learning, vol. 5(3), pp.738-753, 2022.

[40] S. H. I. Nuobei, Q. Zeng, R. Lee, "The design and implementation of language learning chatbot with xai using ontology and transfer learning", In CS & IT Conference Proceedings, vol. 10 (11), Sept. 2020.

[41] M. B. Shabani, "Examining Autonomy, Competence, and Relatedness in CALL: The Case of CALL Apps with the Most Active Users". Computer Assisted Language Learning, vol. 24(1), pp. 222-241, 2023.

[42] S. Symonenko, N. Zaitseva, V. Osadchyi, K. Osadcha, and E. Shmeltser, "Virtual reality in foreign language training at higher educational institutions", CEUR Workshop Proceedings (CEUR-WS.org), paper 3, 2020.

[43] B. Klímová, and P. M. Ibna Seraj, "The use of chatbots in university EFL settings: Research trends and pedagogical implications", Frontiers in Psychology, vol. 14, p. 1146, 2023.

[44] N. Y. Kim, Y. Cha, & H. S. Kim, "Future English learning: Chatbots and artificial intelligence", Multimedia-Assisted Language Learning, vol. 22(3), 2019.

[45] V. Timpe-Laughlin, T. Sydorenko, and P. Daurio, "Using spoken dialogue technology for L2 speaking practice: what do teachers think?", Computer Assisted Language Learning, vol. 35(5-6), pp. 1194-1217, 2022.

[46] M. Brinegar, "Chatbots as supplementary language learning tool: advantages, concerns, and implementation", International Journal of Education and Social Science Research, vol. 6, (6), pp. 223-230, Nov.-Dec. 2023.

2

A Graph Convolutional Network for Channel and Power Allocation in Device-to-device Communication

Lei Gao[1], Xinxin Zhang[2], Shan Yang[1], Xiaoming She[1], Peng Chen[1], and Jianxiu Wang[1]

[1]China Telecom Research Institute, China
[2]School of Information and Communication Engineering, Beijing University of Posts and Telecommunications, China
E-mail: zhangx2@bupt.edu.cn; gaol8@chinatelecom.cn

Abstract

Device-to-device (D2D) communication provides a flexible and efficient way to exchange and deliver information in 5G cellular networks. In real-world scenarios, it is crucial to consider the interference among users and perform proper resource allocation. However, existing schemes often ignore the topological nature of the communication environment and fail to fully leverage its structured information. This chapter investigates the benefits of the topological structure of users in resource allocation tasks and proposes an intelligent allocation model (named CG-GCN) based on full channel gains and a graph convolutional network (GCN). First, an approach is proposed to construct graphs of communication networks, which is to generate edge information of graphs by accumulating channel gains. Then, a GCN-based model for graph data is designed to complete a channel classification task and a transmit power level classification task for all D2D users. Meanwhile, supervised learning combined with unsupervised learning is used to train the model. Numerical results, with the optimization objective of maximizing either spectral efficiency or energy efficiency, validate that the CG-GCN

DOI: 10.1201/9788743801108-3

model approaches the optimal solution and outperforms existing state-of-the-art deep learning models and several traditional schemes.

Keywords: D2D, interference channel, channel allocation, power control, graph convolutional network

2.1 Introduction

Device-to-device (D2D) communication is considered a key technology for 5G cellular networks, allowing D2D pairs in close proximity to each other to exchange data directly without the aid of a base station (BS). In the cellular network where multiple D2D users (DUEs) coexist with cellular users (CUEs), the DUEs can share the radio resources of the CUEs to improve the overall throughput of the communication system. However, due to the existence of multiuser interference among DUEs and interference between DUEs and conventional CUEs, the rational design of resource allocation schemes for D2D scenarios (i.e., determining the channels and transmission power of DUEs) is extremely challenging [1].

Resource allocation for DUEs in cellular networks is usually treated as a nonconvex optimization problem. Accordingly, traditional iterative algorithms [2, 3] suffer from low flexibility, tendency to be trapped in local optimal solutions, and high time complexity in solving such problems. In recent years, with the artificial intelligence (AI) techniques being widely focused on, deep learning (DL) has successfully contributed to advances in wireless communications [4–5] (e.g., being applied to MIMO [6] and mmWave [7] techniques at the physical layer, and intelligent routing schemes at the network layer [8]). For the network of D2D communication, several literatures have designed DL-based resource allocation schemes with different ideas. Based on reinforcement learning, algorithms such as Q-learning [9], DDPG [10], and D3QN [11] were gradually applied to the D2D scenario to formulate optimal resource allocation through the continuous interaction between agents and the wireless environment. Based on unsupervised learning, Lee et al. [12] maximized the total rate of DUEs under a fixed quality of service (QoS) constraint, and a heuristic algorithm combined with a DNN model was designed to control the transmitting power of DUEs assigned to each channel. Simulation results showed that the proposed scheme was close to the optimal sum rate. Further, a DNN model combined with supervised and unsupervised learning was proposed to maximize the SE of DUEs while guaranteeing a minimum spectral efficiency (SE) of CUEs [13]. Experiments revealed that the model determined the channel and transmit power levels of D2D users with high performance and low latency, and outperformed the idea of using only unsupervised learning regarding convergence speed.

Even though deep learning (DL) models have significantly advanced device-to-device (D2D) communication, most existing schemes only incorporate channel state information (CSI) in the D2D environment as an independent dimensional feature for resource allocation modeling. At the model's input stage, these schemes fail to consider the topological relationships among cellular users (CUEs), D2D users (DUEs), and CSI collectively. However, the D2D communication environment, as a naturally structured network, has topological information that should not be neglected. Nevertheless, for graph-structured data with more complex representations, it is one of the challenges to model accurately. Graph neural network (GNN) [14] is a typical model successfully applied to graph data, and its core idea is the message-passing mechanism, which takes each node in the graph to continuously update its state by exchanging information with each other until it converges to a stable state. Subsequently, inspired by the widely used convolutional neural networks [15], by extending convolutional operations from traditional data to graph data, Kipf et al. [16] proposed the graph convolutional network (GCN), which can effectively improve the prediction of graph data and has achieved excellent performance in natural language processing [17], computer vision [18], and other fields.

Motivated by the aforementioned challenges, we design a GCN-based D2D resource allocation model considering full channel gain (named CG-GCN). The main contributions of this chapter are as follows:

- We proposed a method of constructing a graph considering a topology relation of all users, that is, the CSI of multiple channels is used to calculate new edge weights in an accumulative manner, and finally, a directed graph describing the D2D communication environment is generated.
- We design a GCN-based channel allocation and power level classification task, and the training of the model parameters is completed by supervised learning combined with unsupervised learning.
- Simulation experiments show that the results of the proposed CG-GCN model can be extremely close to the optimal allocation scheme and the overall performance is stronger than the recent state-of-the-art scheme (DNN model).
- The proposed scheme is allowed to be applied to different resource allocation objectives (e.g., maximizing energy efficiency) by modifying the loss function, which has a favorable extensibility.

The rest of this chapter is organized as follows. Section 2.2 gives the system model and the problem formulation. Section 2.3 provides a GCN-based scheme. Section 2.4 evaluates the performance of the scheme through several simulation experiments. Finally, Section 2.5 gives the conclusion.

2.2 System Model and Problem Formulation

The system model studied in this chapter is shown in Figure 2.1. In this case, there are K CUEs (K channels, the candidate set $\mathrm{K} = \{1, \ldots, K\}$) and N D2D pairs (each pair consists of a transmitter and receiver, the candidate set $\mathrm{N} = \{1, \ldots, N\}$) in the network with a BS located at the center of the network. To improve the spectrum utilization in the uplink, the DUEs are allowed to multiplex the same channels as the CUEs [13].

In the network, signal fading and interference among signals within the same channel are inevitable. For the kth channel, $h^{\mathrm{k}}_{\mathrm{i,j}}$ represents the channel gain between the transmitter of one D2D pair j to the receiver of another D2D pair i, which is composed of path loss and multipath fading. The QoS of the DUEs is affected due to the existence of CUEs, that is, $h^{\mathrm{k}}_{\mathrm{i,C}}$ represents the channel gain between the CUEs and the receiver of the DUEs. Meanwhile, the signals emitted by the DUEs will affect the communication quality between the BS and the CUEs, that is, $h^{\mathrm{k}}_{\mathrm{C,j}}$ denotes the channel gain between the transmitter of the DUEs and the BS. In addition, $h^{\mathrm{k}}_{\mathrm{C,C}}$ is used to denote the

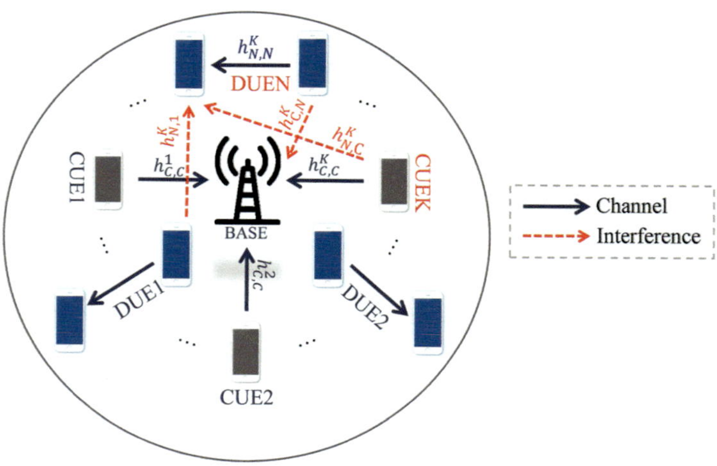

Figure 2.1 System model with three D2D pairs and three CUE users ($K = N = 3$, note: not all links are shown in the figure).

channel gain between the CUEs and the BS. All the channel gains (H) form the full CSI.

$$\begin{aligned} \mathrm{P}: & \max_{f_{\mathrm{a}}^{\mathrm{i,k}}, f_{\mathrm{p}}^{\mathrm{i}}} \xi \left[\sum_{i \in \mathrm{N}} \sum_{k \in \mathrm{K}} \mathrm{SE}_{\mathrm{i}}^{\mathrm{k}} \right], \\ \text{s.t. } C1 & : f_{\mathrm{a}}^{\mathrm{i,k}}(h) \in \{0,1\} \quad \forall i \in \mathrm{N}, \forall k \in \mathrm{K}, \forall h \in \mathrm{H} \\ C2 & : \sum_{\mathrm{k} \in \mathrm{K}} f_{\mathrm{a}}^{\mathrm{i,k}}(h) \leq 1 \quad \forall i \in \mathrm{N}, \forall h \in \mathrm{H} \\ C3 & : f_{\mathrm{p}}^{\mathrm{i}}(h) \in \mathrm{L} \quad \forall i \in \mathrm{N}, \forall h \in \mathrm{H} \\ C4 & : \mathrm{SE}_{\mathrm{C}}^{\mathrm{k}} > SE_{\mathrm{thr}}^{\mathrm{C}} \quad \forall k \in \mathrm{K}, \forall h \in \mathrm{H} \ldots \end{aligned} \quad . \tag{2.1}$$

In this chapter, resource allocation is conducted in a centralized manner, the BS collects the CSI of all users in the environment and generates the resource allocation strategy, which is then distributed to each user. The process for the BS to acquire CSI is as follows: each user transmits a pilot signal in advance, user x estimates $h_{\mathrm{x,y}}$ based on the pilot signal sent by user y, and transmits this information to the BS.

Our optimization objective is to maximize the SE of DUEs when the SE of CUEs is greater than $\mathrm{SE}_{\mathrm{thr}}^{\mathrm{C}}$. Based on all the CSI, the optimal channel $f_{\mathrm{a}}(h)$ and transmission power $f_{\mathrm{p}}(h)$ of DUEs need to be solved. The objective to be optimized is shown in eqn (2.1), where ξ represents the expectation and the SE of DUEs is shown in eqn (2.2), $h_{\mathrm{i,i}}^{\mathrm{k}}$ represents the beneficial channel gain, P_{C} is the transmission power of the CUEs, and $N_0 W$ represents the power of conventional Gaussian white noise. Similarly, the SE of the CUEs is shown in eqn (2.3). Additionally, the optimization objective includes four constraints: C1 and C2 restrict each D2D pair to occupy only one channel; referring to 3GPP [19]; C3 represents the power of DUEs needs to satisfy the discrete class, that is, $P_{\mathrm{i}} \in \mathrm{L} = \{P_1, \ldots, P_8\}$ [13], which is better than the continuous power for practical deployment; C4 requires that the SE of CUEs to exceed a threshold value, that is, to satisfy the minimum communication requirements. The expression of the objective function shows that it is a nonconvex optimization problem with integer constraints and thus has no analytic solution. This motivates an AI-based scheme.

$$\mathrm{SE}_{\mathrm{i}}^{\mathrm{k}} = \log_2 \left(1 + \frac{h_{\mathrm{i,i}}^{\mathrm{k}} f_{\mathrm{a}}^{\mathrm{i,k}}(h) f_{\mathrm{p}}^{\mathrm{i}}(h)}{N_0 W + \sum_{j \in \mathrm{N}(\neq i)} h_{\mathrm{i,j}}^{\mathrm{k}} f_{\mathrm{a}}^{\mathrm{j,k}}(h) f_{\mathrm{p}}^{\mathrm{j}}(h) + h_{\mathrm{i,C}}^{\mathrm{k}} P_{\mathrm{C}}} \right). \tag{2.2}$$

$$\mathrm{SE}_{\mathrm{C}}^{\mathrm{k}} = \log_2\left(1 + \frac{h_{\mathrm{C,C}}^{\mathrm{k}} P_{\mathrm{C}}}{N_0 W + \sum\limits_{j \in \mathrm{N}} h_{\mathrm{C,j}}^{\mathrm{k}} f_{\mathrm{a}}^{\mathrm{j,k}}(h) f_{\mathrm{P}}^{\mathrm{j}}(h)}\right). \tag{2.3}$$

2.3 GCN-based Resource Allocation for D2D

In order to approximate the optimal solution of eqn (2.1), we propose a resource allocation scheme based on GCN. Algorithm 2.1 shows the entire computational process of the scheme. After inputting the relevant parameters, the execution process is mainly divided into constructing graph, modeling, and training, which will eventually obtain the deterministic channels and transmit power levels of DUEs, details of which are shown in subsequent subsections of this chapter, respectively.

2.3.1 Construction graph

The cornerstone of GNN is the construction of graph. Figure 2.2 shows the process of generating a graph from a single sample. Here, each channel can be seen as a directed graph $G = (V, E)$. The graph has $N + 1$ nodes $V = \{v_1, v_2, \ldots, v_{N+1}\}$, including N D2D pairs and one cellular user. Meanwhile,

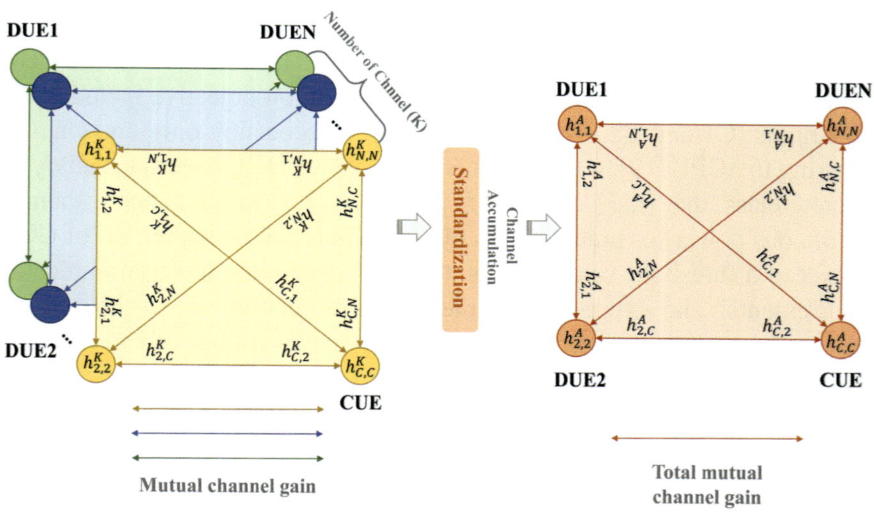

Figure 2.2 Diagram of the construction graph for one sample.

the whole communication environment includes K cellular users. For each node, its feature is the CSI vector after zero-mean normalization. For each sample, according to eqn (2.4), the CSI of K channels is summed to obtain $h_{\mathrm{i,j}}^{\mathrm{A}}$, which is used as the weight of the edges $E = \{v_1 v_1, v_1 v_2, \ldots, v_{N+1} v_{N+1}\}$ among the nodes. Finally, we evaluated the graph construction complexity. In this scenario, the combined feature dimensions of the nodes are $(N+1)^2$. Additionally, each pair of nodes is connected, resulting in an edge weight matrix with dimensions $(N+1)^2$. Therefore, the graph construction complexity of the single sample can be represented as $2\,(N+1)^2$.

$$h_{\mathrm{i,j}}^{\mathrm{A}} = \sum_{k=1}^{K} h_{\mathrm{i,j}}^{\mathrm{k}}. \tag{2.4}$$

2.3.2 CG-GCN model

Further, for the constructed graph, we will build a resource allocation model based on GCN. The information processing for GCN is shown in Figure 2.3. The schematic diagram of the GCN with multiple hidden layers is shown in Figure 2.4. The network has N_x feature dimensions in the input layer and N_y feature dimensions in the output layer. The graphical structure is shared among the layers.

The core idea of GCN is to learn efficient function mapping. Each node aggregates its own features and those of its neighbors through the function mapping to generate a new representation. The update process of the node embedding at the *l*th layer is shown in eqn (2.5), where, the calculation process of $\tilde{A}$ is shown in eqn (2.6), which denotes the summation of the adjacency matrix and the unit matrix of the node. $\tilde{D}$ denotes the degree matrix of $\tilde{A}$. $\tilde{D}^{-0.5}\tilde{A}\tilde{D}^{-0.5}$ is a symmetric normalized Laplacian operation. $H^{(e)}$ is the feature matrix of layer *e*. $W^{(e)}$ is the learnable parameters of layer *e*, which will be updated by the gradient descent algorithm. σ denotes the activation function (default is ReLU).

$$H^{(\mathrm{e}+1)} = \sigma\left(\tilde{D}^{-\frac{1}{2}}\tilde{A}\tilde{D}^{-\frac{1}{2}}H^{(e)}W^{(e)}\right). \tag{2.5}$$

$$\tilde{A} = A + I_n. \tag{2.6}$$

Figure 2.3 includes two graph classification tasks, that is, channel allocation prediction and transmit power level classification. After the data of a batch (consisting of multiple graphs) is input, the GCN is used to generate a new feature matrix based on the mapping function derived from the training.

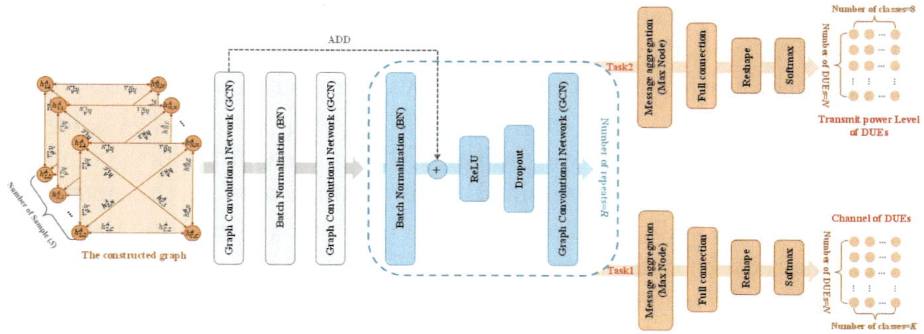

Figure 2.3 Overall diagram of the GCN-based resource allocation model.

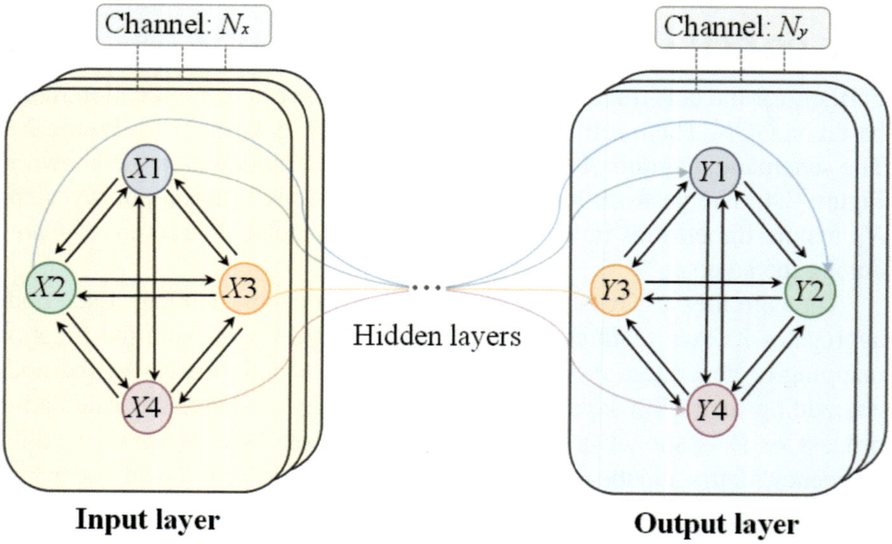

Figure 2.4 The information processing process for GCN.

Then, the data are passed through the batch normalization (BN) module, which executes the normalization operation through learnable parameters (W_{BN} and b_{BN}), mean and variance [20], which effectively avoid overfitting and gradient vanishing phenomena of the model. Further, the data passes through the GCN module and a new feature matrix is generated. Next, a new module with a repeatable number R is designed, passing in turn along the direction of the information flow through the BN operation, a residual connection for promoting training, a ReLU activation function to introduce nonlinearity, a dropout operation for avoiding overfitting and the GCN module for feature generation. Then, the obtained new graph executes the node information aggregation operation which is shown in eqn (2.7). Finally, the information accesses the fully connected layer, and the optimal channel and transmit power levels of the D2D pairs are calculated after the Reshape operation and the Softmax activation operation, respectively.

$$h_{\mathrm{g}} = \max_{v \in V} \{h_{\mathrm{v}}\} . \tag{2.7}$$

Algorithm 2.1. CG-GCN model for resource allocation.

Input: The number of DUEs (N); the number of CUEs (K); CSI for all users; training rounds *R1* and *R2*.

Execute:

Step 1: *construction graph.* $G=(V,E)$, the core elements are as follows:

- ✧ V : N D2D pairs and one cellular user;
- ✧ Feature of V : The CSI vector after zero-mean normalization;
- ✧ E : Cumulative channel gain of each user to other users;
- ✧ Label of G : The target channels and transmit power levels of the DUEs.

Step 2: *modeling.* The channel allocation and power level classification models of all DUEs are built based on GCN respectively.

Step 3: *training.* Include two steps, as follows:

for round from 1 to *R*1 *do*:

Step 3.1: Supervised learning.

end

for round from 1 to *R*2 *do*:

Step 3.2: Unsupervised learning.

end

Output: channels and transmit power levels of the DUEs.

2.3.3 Loss function

To promote better convergence of the CG-GCN model parameters, a combination of supervised and unsupervised learning is used [13]. The loss functions corresponding to the two processes are different.

The loss function for supervised learning is shown in eqn (2.8) and consists of a cross-entropy combined with a binarization term. Where, $a_{\mathrm{i}}^{\mathrm{k}}$ indicates whether the ith D2D pair corresponds to the kth channel, and $p_{\mathrm{i}}^{\mathrm{l}}$ indicates whether the ith D2D pair is the transmit power level l. $\hat{a}_{\mathrm{i}}^{\mathrm{k}}$ and $\hat{p}_{\mathrm{i}}^{\mathrm{l}}$ are the corresponding target values, respectively. They are obtained through an optimal scheme, which is an iterative algorithm that ensures the best combination of channels and power can be achieved. In the binarization term, λ_0 is a control coefficient and g is calculated as shown in eqn (2.9) ($\alpha$is a control coefficient).

$$\mathrm{L}_{\mathrm{SL}} = -\sum_{i\in\mathrm{N}}\left(\sum_{k\in\mathrm{K}} \hat{a}_{\mathrm{i}}^{\mathrm{k}} \log\left(a_{\mathrm{i}}^{\mathrm{k}}\right) + \sum_{l=1}^{8} \hat{p}_{\mathrm{i}}^{\mathrm{l}} \log\left(p_{\mathrm{i}}^{\mathrm{l}}\right)\right) + \lambda_0 g. \tag{2.8}$$

$$g = -\sum_{i\in\mathrm{N}}\left(\sum_{k\in\mathrm{K}} \left|a_{\mathrm{i}}^{\mathrm{k}} - 0.5\right|^{\alpha} + \sum_{l=1}^{8} \left|p_{\mathrm{i}}^{\mathrm{l}} - 0.5\right|^{\alpha}\right). \tag{2.9}$$

The unsupervised learning will continue the training for the parameters obtained from the supervised learning, and the loss function is shown in eqn (2.10), which consists of an optimization objective, a constraint term, and a binarization term, where, λ_1 and λ_2 are the new control coefficients, and δ represents a minimal value tending toward 0.

$$\mathrm{L}_{\mathrm{USL}} = -\sum_{i\in\mathrm{N},k\in\mathrm{K}} \mathrm{SE}_{\mathrm{i}}^{\mathrm{k}} + \lambda_1 \sum_{k\in\mathrm{K}} \frac{\max\left(\mathrm{SE}_{\mathrm{thr}}^{\mathrm{C}} - \mathrm{SE}_{\mathrm{C}}^{k}\right)}{\mathrm{SE}_{\mathrm{thr}}^{\mathrm{C}} + \delta} + \lambda_2 g. \tag{2.10}$$

2.4 Performance Evaluation

2.4.1 Simulation environment and parameter setting

We will evaluate the performance of the proposed model in a D2D environment through several simulation experiments. The study uses Python 3 and Torch 1.9.0 to implement the proposed algorithm. The relevant computations for the experiments were all performed on a GPU (NVIDIA TITAN Xp) device. To facilitate comparison with recent excellent models (based on DNN), the parameters of the wireless environment are set with reference to [13], $N = K = 3$, the area of the region is 100 m $\times$ 100 m, the maximum distance of the D2D pair is 30 m, $\mathrm{SE}_{\mathrm{thr}}^{\mathrm{C}} = 0$ (corresponding to low-rate scenarios for cellular users), and other parameters of the wireless environment are shown in Table 2.1.

Table 2.1 Simulation parameters setting for D2D scenario.

Parameters (D2D scenario)	**Values**
W	100 MHz
N_0	−173 dBm/Hz
$P_1 \sim P_8$ (mW)	0, 28.57, 57.14, 85.71, 114.28, 142.85, 171.42, 200
λ_0	1
λ_1	70
λ_2	15
α	4

In addition, the parameters of the CG-GCN model directly affect the prediction performance. After several debugging experiments, the parameters are set as follows: Adam is used as the optimizer; the learning rates supervised and unsupervised learning are set to 0.001 and 0.000001, respectively, and the number of steps is set to $R1 = 7$ and $R2 = 200$, respectively; the number of repeats of submodules is $R = 6$; and the number of nodes in the hidden layer is 400.

2.4.2 Performance evaluation for supervised learning

To evaluate the performance of the model under supervised learning, following [13], we randomly generate 70,000 samples as required by the D2D communication environment and divide them into the training set, development set (DEV), and test set (TEST) in 6: 2: 2. Meanwhile, the sample labels are generated using an exhaustive method, which ensures the optimal resource allocation results are obtained.

The error variation of the training process is shown in Figure 2.5. The error of the CG-GCN model is consistently lower than the DNN model from the beginning, demonstrating a better parameter training capability. Moreover, we evaluated the accuracy variation of the development set during the training process for the resource allocation task. The result of the channel classification task is shown in Figure 2.6, where the accuracy of the CG-GCN model is consistently higher than that of the DNN model, with a final difference of 1.54%. In Figure 2.6, the transmit power level classification task also shows a similar trend, but the accuracy of the DNN model has a significant downward trend after 40 steps, which indicates that there is a high probability of overfitting from this point on. In contrast, the performance of the CG-GCN model keeps increasing and ends up with an accuracy 4.39% higher than the DNN.

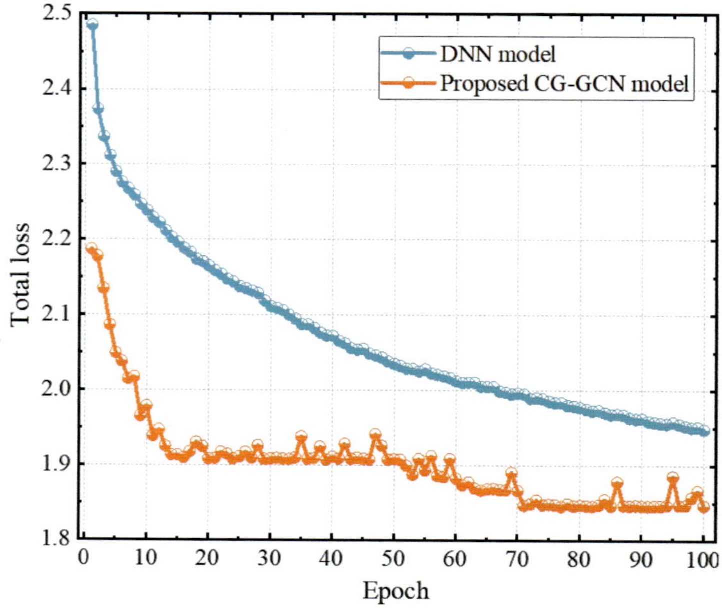

Figure 2.5 Error variation during model training.

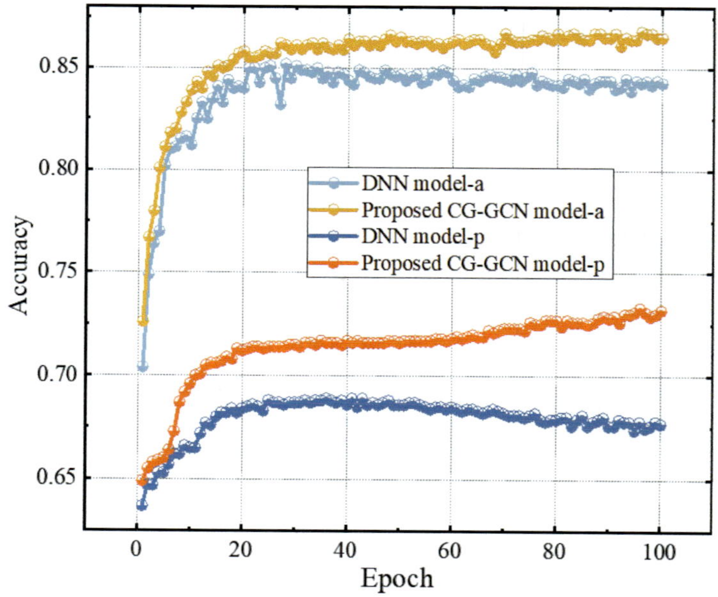

Figure 2.6 Accuracy variation of the channel classification task and the transmit power level classification task for the development set during training.

Table 2.2 Accuracy of models under supervised learning.

Model	DEV (%)		TEST (%)	
	$f_a(h)$	$f_p(h)$	$f_a(h)$	$f_p(h)$
DNN [13]	85.27	68.86	84.99	68.91
CG-GCN (R = 6)	**86.81**	**73.25**	**86.50**	**73.30**
CG-GCN (R = 16)	**87.00**	**73.90**	**86.89**	**74.21**

Table 2.3 The number of parameters for different models.

Model	The number of parameters θ
DNN [13]	5.5 M
CG-GCN (R = 6)	**2.3 M**
CG-GCN(R = 16)	5.4 M

After the parameters of the model are trained, the test set is used to eventually evaluate the performance of the model. The results are shown in Table 2.2. For the channel allocation task, the accuracy of the CG-GCN model is 1.51% higher than the DNN model, reaching 86.50%. For the transmit power level classification task, the accuracy is only 73.30%, which is due to the higher complexity of this task compared to the channel allocation and more classes need to be classified. However, the accuracy of the CG-GCN model is 4.39% higher than the DNN. This demonstrates that our scheme has a more advanced performance for complex power allocation tasks.

The overall performance of the proposed model is influenced by the value of R, as it consists of modular components. When $R = 16$, the model demonstrates better performance, with a slight improvement in accuracy on both the development and test sets. However, as shown in Table 2.3, the model parameters double compared to when $R = 6$, which is not conducive to practical deployment. Considering the performance comparison of the models in Table 2.2, the CG-GCN with $R = 6$ is optimal.

2.4.3 Overall performance evaluation of the model

To perform an overall evaluation of the model, 1,000,000 samples are randomly generated in the D2D communication environment for the unsupervised learning of all model parameters, where the first 50,000 samples are used to perform the supervised learning. After that, 100,000 samples are randomly generated as a test set. To visualize the performance gap between the proposed model and the optimal scheme (exhaustive search), we plot the cumulative distribution functions (CDF) of SE for DUEs and CUEs, respectively, as shown in Figures 2.7 and 2.8, where, the random scheme refers to the random selection of channels and transmission power levels.

For the CDF of DUEs, it can be observed that the distance between the intelligent algorithm and the optimal solution is greatly reduced compared

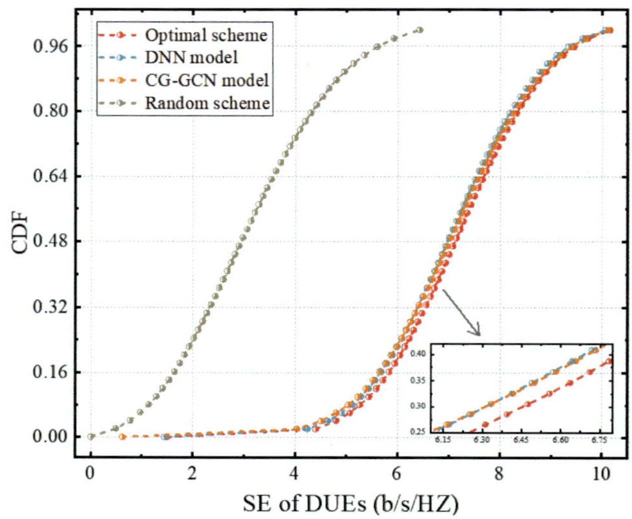

Figure 2.7 CDF of SE for DUEs.

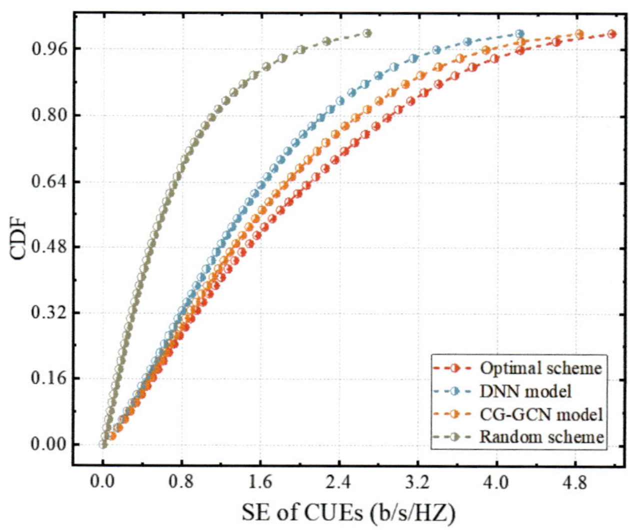

Figure 2.8 CDF of SE for CUEs.

to the random scheme, which indicates the effectiveness of the DL-based model. It is worth noting that both the CG-GCN model and the DNN are extremely close to the curve of the optimal scheme, and the performance of CG-GCN is slightly better than the DNN. Regarding the CDF of CUEs, the CG-GCN model significantly outperforms DNN and is closer to the value of the optimal scheme. Meanwhile, Lee et al. [13] have demonstrated that the DNN model outperforms other traditional heuristic methods (e.g., the equally reduced power (ERP) scheme and the interference pricing (IP) scheme [21]). This shows that our model further enlarges the performance gap between DL-based and traditional schemes.

2.4.4 Performance evaluation of model with maximizing EE

To validate the strong extension ability of the proposed model, we change the optimization objective of the D2D communication system from maximizing SE to maximizing energy efficiency (EE). EE is an important metric for assessing communication system performance, and its conversion formula is shown in eqn (2.11), where p_{i} represents the transmission power of D2D users and $\hat{p}$ is the circuit power (default is 500 mW). From eqn (2.11), it can be inferred that an increase in power will lead to a simultaneous increase in both the numerator and denominator of EE, thus resulting in a higher complexity. Additionally, the configuration of the data set scale, CG-GCN model structure, and error propagation method remained the same as in the SE experiment.

$$\mathrm{EE}_{\mathrm{i}}^{\mathrm{k}} = \frac{\mathrm{SE}_{\mathrm{i}}^{\mathrm{k}}}{p_{\mathrm{i}} + \hat{p}}. \tag{2.11}$$

The CDF of DUEs and CUEs for the test set is shown in Figure 2.9. As observed from the figure, overall, DL-based approaches are far superior to random approaches, with the CG-GCN model outperforming the DNN model and approaching the optimal solution. Combining with the [13], the DNN model demonstrated in maximizing EE exceeded heuristic schemes ERP and IP, further highlighting the effectiveness of the proposed CG-GCN model.

2.4.5 Measurement of computational efficiency for model

Finally, we recorded the time taken by several schemes to perform resource allocation on the test set, which is a necessary consideration in practical deployment scenarios. The results are shown in Table 2.4. Both the CG-GCN and DNN effectively reduce the time cost compared to the optimal scheme.

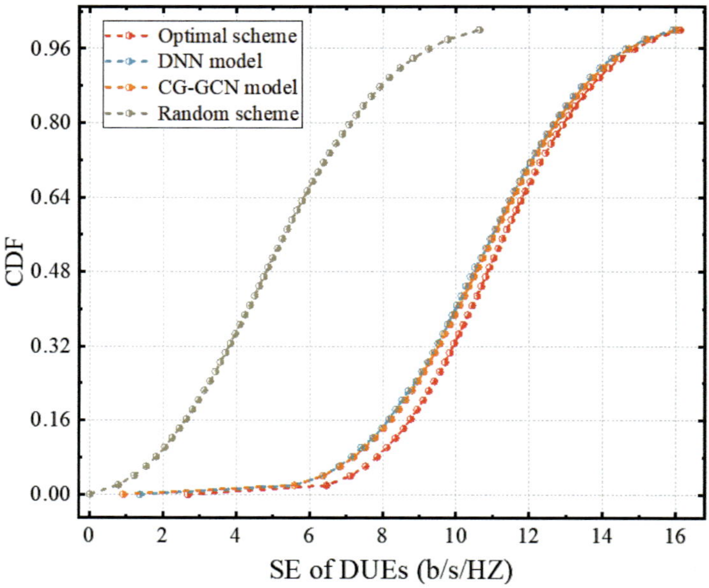

(a)

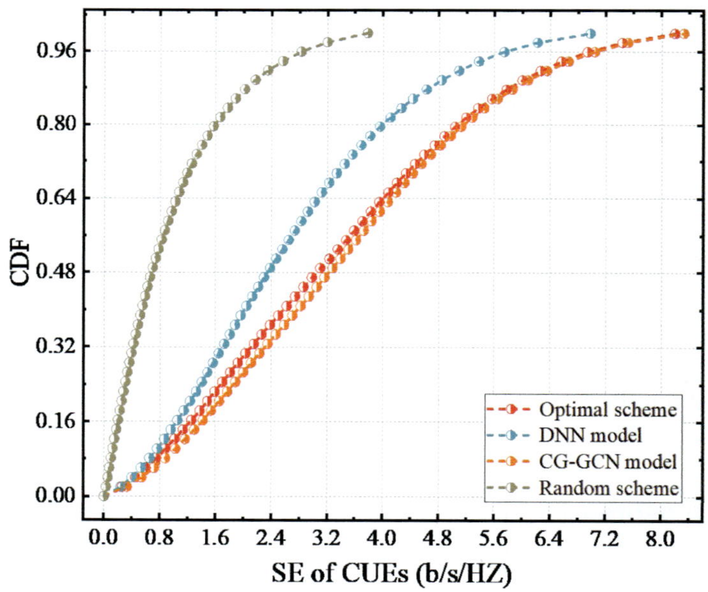

(b)

Figure 2.9 CDF of EE for DUEs (a) and CUEs (b).

Table 2.4 Calculation time(s) for different schemes.

	Optimal scheme	**DNN scheme**	**CG-GCN scheme**	**Random scheme**
SE	3225.93	63.67	**28.28**	77.24
EE	3230.39	65.89	**30.45**	78.34

Among them, CG-GCN takes only about 30 to compute 100,000 samples, which is about 55% better than the calculation efficiency of the existing optimal DNN model.

2.5 Conclusions

This chapter addresses the often ignored network topology in existing DL-based resource allocation schemes for D2D scenarios and optimizes the channel classification task and transmission power level classification task. First, a method for constructing graphs of network topological relations is developed to generate a directed graph of the communication network by accumulating the channel gains. Then, to perform the whole graph classification task, we propose a GCN-based model to approximate the optimal resource allocation scheme. Finally, the simulation results demonstrate that the proposed model outperforms the DNN model and the traditional schemes in terms of accuracy, complexity, and time cost. At the same time, our model has been proven to adapt to different optimization objectives. Next work, we will explore more complex scenarios, such as full-duplex and RIS-assisted D2D communication, by leveraging deep-level topological information in communication networks to optimize the relevant resource allocation schemes

References

[1] F. Jameel, Z. Hamid, F. Jabeen, S. Zeadally, and M. A. Javed, "A survey of device-to-device communications: Research issues and challenges," IEEE Commun. Surveys Tuts., vol. 20, pp. 2133–2168, 2018.

[2] Y. Yuan, T. Yang, H. Feng, and B. Hu, "An iterative matching-Stackelberg game model for channel-power allocation in D2D underlaid cellular networks," IEEE Trans. Wireless Commun., vol. 17, no. 11, pp. 7456–7471, 2018.

[3] M. Liu and L. Zhang, "Resource allocation for D2D underlay communications with proportional fairness using iterative-based approach," IEEE Access, vol. 8, pp. 143787–143801, 2020.

[4] J. Jiao, X. Sun, L. Fang, and J. Lyu, “An overview of wireless communication technology using deep learning,” China Commun., vol. 18, no. 12, pp. 1-36, 2021.
[5] H. Huang, S. Guo, G. Gui, Z. Yang, J. Zhang, H. Sari, and F. Adachi, “Deep learning for physical-layer 5G wireless techniques: Opportunities, challenges and solutions,” IEEE Wireless Commun., vol. 27, no. 1, pp. 214-222, 2020.
[6] C. Wen, W. Shih, and S. Jin, “Deep Learning for Massive MIMO CSI Feedback,” IEEE Wireless Commun. Lett., vol. 7, no. 5, pp. 748-51, 2018.
[7] H. Huang, Y. Song, J. Yang, G. Gui, and F. Adachi, “Deep-Learning-based Millimeter-Wave, Massive MIMO for Hybrid Precoding,” IEEE Trans. Veh. Technol., vol. 68, no. 3, pp. 3027-32, 2019.
[8] F. Hu, B. Chen, D. Shi, X. Zhang, and H. Z. Pan, “Secure Routing Protocol in Wireless Ad Hoc Networks via Deep Learning,” in 2020 IEEE Wireless Communications and Networking Conference (WCNC), 2020, pp. 1-6.
[9] Q. Guo, F. X. Tang, and N. Kato, “Federated Reinforcement Learning-Based Resource Allocation in D2D-Enabled 6G,” IEEE Network, 2022.
[10] C. Kai, X. Meng, L. Mei, and W. Huang, “Deep Reinforcement Learning Based User Association and Resource Allocation for D2D-enabled Wireless Networks,” in 2021 IEEE/CIC International Conference on Communications in China (ICCC), 2021, pp. 1172-1177.
[11] D. Wang, Q. Liu, J. Tian, Y. Zhi, J. Qiao, and J. Bian, “Deep Reinforcement Learning for Caching in D2D-Enabled UAV-Relaying Networks,” in 2021 IEEE/CIC International Conference on Communications in China (ICCC), 2021, pp. 635-640.
[12] W. Lee and K. Lee, “Resource Allocation Scheme for Guarantee of QoS in D2D Communications Using Deep Neural Network,” IEEE Commun. Lett., vol. 25, no. 3, pp. 887-891, 2021.
[13] W. Lee and R. Schober, “Deep learning-based resource allocation for device-to-device communication.” IEEE Trans. Wireless Commun., vol. 21, no. 7, pp. 5235-5250, 2022.
[14] F. Scarselli, M. Gori, A. C. Tsoi, M. Hagenbuchner, and G. Monfardini, “The Graph Neural Network Model,” IEEE Trans. Neural Netw., vol. 20, no. 1, pp. 61-80, 2009.
[15] Z. Li, F. Liu, W. Yang, S. Peng, and J. Zhou, “A Survey of Convolutional Neural Networks: Analysis, Applications, and Prospects,” IEEE Trans. Neural Netw. Learn., vol. 33, no. 12, pp. 6999-7019, 2022.

[16] T. N. Kipf and M. Welling, "Semi-Supervised Classification with Graph Convolutional Networks," In 5th International Conference on Learning Representations (ICLR), 2017.

[17] H. Ren, W. Lu, Y. Xiao, X. Chang, X. Wang, Z. Dong, and D. Fang, "Graph convolutional networks in language and vision: A survey," Knowledge-Based Syst., vol. 251, 2022.

[18] P. Cao, Z. Zhu, Z. Wang, Y. Zhu, and Q. Niu, "Applications of graph convolutional networks in computer vision," Neural Comput. Appl., vol. 34, no.16, pp. 13387-13405, 2022.

[19] H. H. Nguyen and W. J. Hwang, "Distributed scheduling and discrete power control for energy efficiency in multi-cell networks," IEEE Commun. Lett., vol. 19, no. 12, pp. 2198-2201, 2015.

[20] I. Goodfellow, Y. Bengio, and A. Courville, Deep Learning. Cambridge, MA, USA: MIT Press, 2016.

[21] W. Lee, T. W. Ban, and B. C. Jung, "Distributed Transmit Power Optimization for Device-to-Device Communications Underlying Cellular Networks," IEEE Access, vol. 7, pp. 87617-87633, 2019.

3

A Two-stage Beam Selection Method using Machine Learning and a Peak Finding Algorithm

Haesik Kim

VTT Technical Research Centre of Finland, Finland
E-mail: haesik.kim@vtt.fi

Abstract

The multiple-input and multiple-output (MIMO) techniques have evolved significantly in recent decades. We expect to reshape the 6G cellular networks by adopting new innovative techniques. In particular, beamforming is used to allow for directional multiple beams by weighting the magnitude or phase of signals in multiple antennas. It can improve spectral efficiency, reduce interferences, and expand coverage. It will be helpful for maintaining faster and more reliable connectivity in 6G networks. However, there are still many research challenges to design and implement beamforming systems. We need to optimize beam management schemes in terms of beam selection, connection set-up time, energy efficiency, throughput, network outage, QoS/QoE, and so on. It is not easy to find optimal network parameters and keep good link quality. The 3rd generation partnership project (3GPP) 5G new radio (NR) beam selection is based on an exhaustive search method, grid-of-beams (GoB), and precoding matrix indicator (PMI). The exhaustive search method results in a long beam search time, a high overhead cost, and a high energy consumption. In this chapter, we propose a two-stage beam selection method consisting of machine learning and peak finding algorithm to reduce beam sweeping overhead by predicting the beam selection subset and finding the optimal beam pair efficiently.

Keywords: MIMO, 5G, 6G, beamforming, grid-of-beams, machine learning, deep learning, peak finding algorithm

DOI: 10.1201/9788743801108-4

3.1 Introduction

The multiple-input and multiple-output (MIMO) techniques are now widely used in cellular systems and have gradually evolved in recent decades, which are expected to reshape the 6G wireless communications and networks. In particular, a base station (or gNodeB, gNB) of 5G new radio (NR) networks is equipped with a very large number of antennas. They improve the network performances, as well as increase the coverage. We call them massive MIMO (mMIMO). The mMIMO will be a key technique in 6G systems. The key challenges to implementing mMIMO are to obtain accurate channel state information, reduce overheads, satisfy different variants, and so on. In the case of multiuser MIMO, a base station equipped with a larger number of antennas than mobile stations (or user equipment, UE) transmits multiple data streams to mobile stations at the same time. Thus, we can achieve a high throughput. In 5G, beamforming is used to allow for directional multiple beams by weighting the magnitude or phase of signals in multiple antennas. It can improve spectral efficiency and reduce interferences. Thus, mobile users can receive strong signals and it is helpful for maintaining faster and more reliable connectivity in 5G systems. In particular, it provides us with better coverage for an edge user. There are three types of beamforming schemes: digital beamforming, analog beamforming, and hybrid beamforming. Digital beamforming uses precoding in the baseband domain and designs different signals for antennas. Typically, the multiple beams are created by matrix-based operations including weighted amplitudes and phases. It allows us to control multiple beams without a hardware change efficiently and flexibly. In order to implement digital beamforming, a separate radio frequency (RF) chain per antenna is assumed. It causes a higher complexity. The other disadvantages are high direct current (DC) power, loss of signal, and so on. Analog beamforming adjusts the phases of each antenna in the RF domain and creates an antenna pattern with a specific direction. It brings an impact on the radiation pattern and gain of the antenna. Thus, it improves the coverage, as well as creates narrower beamwidths with simple hardware. However, the key disadvantage is the limited number of beams and not flexible beam generation. Hybrid beamforming (also known as hybrid precoding) combines them to balance computational complexity and performance. Coarse beamforming is performed in the analog domain and then the analog beams adopt a MIMO scheme in the digital domain. It allows us to reduce the number of RF chains and improve the flexibility. It is considered for the 5G NR mmWAVE frequency range. Many research groups now pay attention to developing beamforming techniques and efficient beam selection techniques for 6G networks.

In [1], the 3rd generation partnership project (3GPP) defined the beam management procedure for 5G NR for ideal mode and connected mode. The ideal mode is selected when UE does not have active data transmission. The connected mode is used when active data exchange is taking place. The Layer 1 (L1) and Layer 2 (L2) beam management procedure to obtain an optimal beam pair (where a beam pair means a transmit beam and a corresponding receive beam to establish a connection) can be summarized as follows: Step1 is to acquire frame synchronization and find the cell identity. Step 2 is a beam sweep and measurement. In this step, the gNB periodically transmits beams in all directions. The UE detects all beams in a search area and measures the strength of the beams in terms of reference signal received power (RSRP), signal-to-interference-plus-noise ratio (SINR), or reference signal received quality (RSRQ). Step 3 is beam selection and reporting. In this step, the UE selects the beam with the strongest signal, reports it to the gNB, and establishes connectivity. Each cell includes multiple static synchronization signal block (SSB) beams always pointing in the same direction. They form a grid-of-beams (GoB) and cover the whole cell area [2, 3, 4]. Different SSBs are transmitted at different time slots in a cell. They minimize intracell interferences among SSBs. Thus, the 5G coverage is based on not cells but beams. In order to identify separating beams, physical cell ID (PCI) and beam ID are used. SSB beams are regarded as a new layer of minicells in each cell.

We mentioned many benefits of MIMO and beamforming techniques but there are still many research challenges to design and implement them. In particular, we need to optimize beam management schemes in terms of beam selection, connection set-up time, energy efficiency, throughput, network outage, QoS/QoE, and so on. It is not easy to find optimal network parameters and keep good link quality. Based on the GoB approach of 3GPP, the UEs do not observe actual channels but low-dimensional effective channels. The orthogonal reference signal (RS) is assigned to the beam in the GoB codebook [5]. The performance of the GoB codebook design is relevant to the reporting of the UE. It is directly related to the precoder selection. Thus, we can reduce the overhead by exploiting effective channels but the overhead reduction causes performance degradation because the reduced channel representations might not include the key characteristics of the real-world channels. In [6], an iterative scheme for beam management is proposed. They removed some beams contributing less to the sum rate of a multiuser MIMO. In [7], the authors proposed a low-complexity beam selection using an approximation of the sum rate computed out of the elements of the effective channel matrix at the output of the analog beamforming block. Their approach shows us a near-optimal in the high SNR. The beam training can be divided into initial access and beam tracking. The initial access is to establish an initial connectivity in the absence of channel information. It is required as the first step of beamforming schemes. The beam tracking is performed when connectivity is maintained. It tracks the deviation of the paths and updates the quality of connectivity. The conventional approach is based on exhaustive search algorithms and codebook designs. There are many algorithms to reduce the training complexity and the overhead cost. In [8], the authors proposed multiresolution codebooks to reduce the setup time of beamforming. In [9], a low-complexity hierarchical codebook-based beam alignment is proposed. In [10], a low-complexity primary-auxiliary codebook-based beam training is proposed. As AI techniques are adopted in many research fields, they will be one of the key technologies for 6G. They as game-changers will play a critical role in 6G and rebuild cellular networks. In order to adopt AI techniques to 6G, it is important to collect and store enough training data. Fortunately, 6G would be a good environment to acquire big data and operate AI models. Thus, many research groups are exploring AI techniques and adopting them to wireless systems including beamforming [11, 12]. In order to improve the accuracy of beamforming, extrinsic information such as GPS and Lidar signals is sometimes used for deep learning training [11, 12].

The main contributions of this chapter can be summarized as follows:

(1) A new beam selection method using machine learning and peak finding algorithm is proposed to obtain the optimal beam pair and reduce the beam sweeping overhead.
(2) The performance of the proposed beam selection method is evaluated and analysis and lessons learned are described.

The remaining parts of this chapter are as follows: In Section 3.2, the system model is described. In Section 3.3, a new beam selection method is proposed. In Section 3.4, numerical analysis is introduced, and analysis and lessons learned are included. Section 3.5 contains the conclusion and summary.

3.2 System Model

We consider a single-cell OFDMA-based downlink model. Figure 3.1 illustrates an example of a single-cell downlink beamforming scenario.

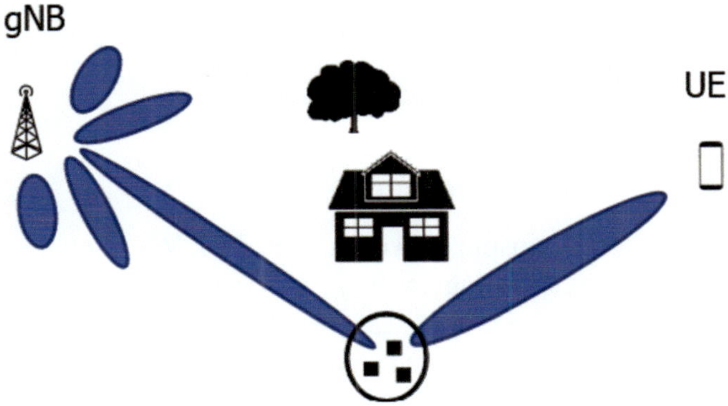

Figure 3.1 Example of single cell downlink beamforming scenario.

The gNB is equipped with $N_t \gg 1$ antennas and has codebooks.

$$B_{gNB} = \left\{ \mathbf{v}_1, \ldots, \mathbf{v}_{|B_{gNB}|} \mid \mathbf{v}_i \in \mathbb{C}^{N_t \times 1} \right\}, \tag{3.1}$$

where $\mathbf{v}_i$ denotes i th beamforming vector in B_{gNB}. The gNB serves $K \ll N_t$ UEs with N_r antennas and codebooks.

$$B_{UE} = \left\{ \mathbf{w}_1, \ldots, \mathbf{w}_{|B_{UE}|} \mid \mathbf{w}_j \in \mathbb{C}^{N_r \times 1} \right\} \tag{3.2}$$

where $\mathbf{w}_j$ denotes j th beamforming vector in B_{UE}. The channel matrix of the k th UE is $\mathbf{H}_k \in \mathbb{C}^{N_r \times N_t}$. We assume the grid-of-beam of 3GPP [1, 3]. The objective of the beam selection is

$$\max_{(i,j) \in \left|B_{gNB}\right| \times |B_{UE}|} g(i,j) = \left\| \mathbf{w}_j^H \mathbf{H}_k \mathbf{v}_i \right\|. \tag{3.3}$$

Typically, we can find the solution by sequential exhaustive search method based on the estimate of $\mathbf{H}_k$. The 3GPP adopted beam management based on this approach. The overhead of beam alignment is proportional to the channel gain uncertainty. In particular, when we deal with vehicular communications or unmanned aerial vehicle (UAV) communications, the channel coherence time is short and the longer overhead of beam alignment is required. Now, we allocate the resource elements (RE) of the resource grid to reference signals (RS) and data. We define the orthogonal frequency-division multiplexing (OFDM) symbol duration τ and the number of the beam training overhead Δ_{gNB} in $B_{gNB}.\tau\Delta_{gNB}$ and RE $\tau\Delta_{gNB}$ are assigned to RS and data, respectively. When the k th UE has the number of activated beam Δ_{UE}, the received beam training overhead signal of the k th UE $\mathbf{Y}_k \in \mathbb{C}^{\Delta_{UE} \times \tau}$ can be expressed as follows: follows:

$$\mathbf{Y}_k = \sqrt{\frac{P}{RE}} \mathbf{W}_k^H \mathbf{H}_k \mathbf{V} \mathbf{R} + \mathbf{W}_k^H \mathbf{N} \tag{3.4}$$

where P is the total transmit power at the gNB, $\mathbf{W}_k$ includes the beamforming vectors corresponding to B_{UE} and represents the k th UE beam training overhead combiner, $\mathbf{V} \in \mathbb{C}^{N_{gNB} \times \Delta_{gNB}}$ includes the beamforming vectors corresponding to B_{gNB} and represents the normalized beam training overhead precoder, $\mathbf{H}_k$ is channel matrix between the gNB and the k th UE, $\mathbf{R} \in \mathbb{C}^{\Delta_{gNB} \times \tau}$ includes the known RSs and has the property $\mathbf{R}\mathbf{R}^H = \mathbf{I}_{\Delta_{gNB}}$, and $\mathbf{N} \in \mathbb{C}^{N_{UE} \times \tau}$ is Gaussian noise. The k th RSs and has the property $\mathbf{R}\mathbf{R}^H = \mathbf{I}_{\Delta_{gNB}}$, and $\mathbf{N} \in \mathbb{C}^{N_{UE} \times \tau}$ is Gaussian noise. The k th UE can estimate the effective channels matrix $\mathbf{H}_k \in \mathbb{C}^{\Delta_{UE} \times \Delta_{gNB}} = \mathbf{W}_k^H \mathbf{H}_k \mathbf{V}$ and report it to the gNB . The effective channel matrix is related to the active beam Δ_{UE}. We assume that each UE has a single data stream, and the transmitted data vector is expressed as $\mathbf{x} \in \mathbb{C}^{K \times 1} = [x_1, \ldots, x_K]$ with $E\left[\mathbf{x}\mathbf{x}^H\right] = \mathbf{I}_K$. The received data signal of the k th UE y_k can be expressed as follows:

$$y_k = \sqrt{\frac{P}{RE}} \overline{\mathbf{w}}_k^H \overline{\mathbf{H}}_k \overline{\mathbf{V}} \mathbf{x} + \overline{\mathbf{w}}_k^H \overline{\mathbf{n}}_k \tag{3.5}$$

where $\overline{\mathbf{w}}_k \in \mathbb{C}^{\Delta UE \times 1}$ is the data combiner, $\overline{\mathbf{H}}_k$ is the effective channel matrix, $\overline{\mathbf{V}} \in \mathbb{C}^{\Delta_{gNB} \times K}$ is the normalized data precoder, and $\overline{\mathbf{n}}_k \in \mathbb{C}^{\Delta_{UE} \times 1} = \mathbf{W}_k^H \mathbf{n}_k$.

The conventional beam training method of 3GPP [1, 3, 4] searches all beams as $B_{gNB} = \Delta_{gNB}$. In order to reduce the overhead, we need to search only the relevant channel components and define the set of the relevant beam training overhead D_k for the k th UE in B_{gNB} as follows:

$$\mathrm{D}_k = \left\{(i, j) \mid E_{\mathbf{H}_k}\left[\left|\mathbf{w}_j^H \mathbf{H}_k \mathbf{v}_i\right|^2\right] \geq \theta\right\} \tag{3.6}$$

where θ is a threshold of the power. D_k contains the relevant beam pairs. In [13], the relationship between beam coherence time and channel coherence time is investigated. The channel coherence time T_C is already well defined as the time over which the channel response does not vary. The beam coherence time T_B is defined as the average period over which the beams are unchanged. For a given receive beam, when the received power $P(t)$ is below a ratio $\xi \in$

$[0, 1]$ compared to the peak receive power, the receive beam points at the peak direction μ_r at time t and the beam coherent time is defined as follows [13]:

$$T_B = \inf_{\gamma} \left\{ \gamma \left| \frac{P(t+\gamma)}{P(t)} < \xi \right. \right\}, \tag{3.7}$$

and the power decrease is caused by the pointing error. D_k only relies on the second order statistics (correlation properties of wireless channels) of the channel matrix $\mathbf{H}_k$. The beam coherence time is much longer than the channel coherence time [14]. When the *k*th UE beam training overhead combiner $\mathbf{W}_k$ is selected, we can represent the beam selection subset as $\mathrm{D}_k(\mathbf{W}_k)$ that can be regarded as a precoding matrix indicator (PMI). In 3GPP [1, 3, 4], the UE reports the desired PMI to the gNB. The PMI report is about what precoding matrix is used for the downlink channel. More specifically, the PMI report is not a matrix but coefficients. The gNB finds weights for beamforming and constructs a precoding matrix using the coefficients. The combination of them is limited. Likewise, the *k*th UE has the second order statistics and report $\mathrm{D}_k(\mathbf{W}_k)$ to the gNB when conventional exhaustive search method is used. In this chapter, we assume that $\mathrm{D}_k(\mathbf{W}_k)$ is not specific coefficients but the beam selection subset information. The UEs do not report the PMI and the gNB finds the relevant searching sets in all candidate sets $\bigcup_{k=1}^{K} \mathrm{D}_k\left(\mathbf{W}_k\right)$ using machine learning algorithms. We target to minimize the overhead and find the optimal beam pairs.

There are many ways to design the precoders such as minimum mean square error (MMSE), matched filtering (MF), and zero forcing (ZF). The ZF precoder as one of the popular techniques creates orthogonal channels for each UE and applies a transformation matrix to the transmit signal. We assume that the gNB uses the ZF precoder and UE has a single receive beam such as $\Delta_{UE} = 1$. Thus, we can simplify the effective channel matrix $\overline{\mathbf{H}}_k \in \mathbb{C}^{K \times \Delta_{gNB}}$ between the gNB and the k th UE as follows:

$$\overline{\mathbf{H}}_k = \left[\overline{\mathbf{h}}_1^T, \ldots, \overline{\mathbf{h}}_K^T\right]^T \tag{3.8}$$

where $\overline{\mathbf{h}}_k = \mathbf{w}_k^H \mathbf{H}_k \boldsymbol{V}$. We can rewrite eqn (3.5) as follows:

$$y_k = \sqrt{\frac{P}{RE}} \rho \overline{\mathbf{H}}_k \overline{\mathbf{V}}_{ZF} \mathbf{x} + \mathbf{n} \tag{3.9}$$

where $\overline{\mathbf{V}}_{ZF} \in \mathbb{C}^{\Delta_{gNB} \times K}$ is ZF precoder and ρ is the power normalization factor. Now, we can define the optimal beam allocation at the gNB using the maximum sum rate criterion [14] as follows:

$$g^{*}\left(\mathbf{w}_{k}\right)=\underset{\mathbf{w}_{k}}{\operatorname{argmax}} \sum_{k=1}^{K} \log_{2}\left(1+\zeta_{k, \mathbf{w}_{k}}\right) \tag{3.10}$$

where ζ_{k,W_k} denotes the beam strength at the k th UE for the beam selection $\mathbf{w}_k$. The beam strength can be RSRP, RSRQ, or SINR. If the channel state information is fully known, the optimization problem eqn (3.10) can be solved by an exhaustive search method with the complexity $O(\mathbf{W}_k)$. The complexity increases exponentially as the number of user K increases.

3.3 Proposed Two-stage Beam Selection Method

In Section 3.2, we constructed the system model based on the GoB and PMI of 3GPP 5G NR. In this section, we propose a two-stage beam selection method consisting of machine learning and peak finding algorithm to reduce beam sweeping overhead.

3.3.1 Outline for two-stage beam selection method

The beam selection of 5G NR is based on the exhaustive search method systematically checking all possible candidates whether or not they satisfy the statement of the problem. The beam selection subset $\mathrm{D}_k(\mathbf{W}_k)$ as a PMI is regularly reported to the gNB. The exhaustive search method results in a long beam search time, a high overhead cost, and a high energy consumption. In this chapter, we improve the beam selection process using machine learning and peak finding algorithm. In the first stage, we find the beam selection subset $\mathrm{D}_k(\mathbf{W}_k)$ using machine learning from a history data set. Therefore, the UEs do not report all of the PMI information regularly. The overhead cost can be reduced. In the second stage, we search the optimal beam pair from the beam selection subset $\mathrm{D}_k(\mathbf{W}_k)$ by the peak finding algorithm. The proposed two-stage beam selection algorithm is described in the next sections.

3.3.2 Beam selection subset using machine learning

The optimal beam allocation problem eqn (3.10) can be regarded as classification problem of machine learning. We can have a labeled data set as a pair of $(\mathbf{w}_k, g^*(\mathbf{w}_k))$ and construct a neural network consisting of the input nodes of $\mathbf{w}_k$ and the output nodes of $g^*(\mathbf{w}_k)$. We find the optimal beam allocation

$g^*(\mathbf{w}_k)$ for the given input $\mathbf{w}_k$ that is equivalent to solve the problem eqn (3.10). The beam selection is mapped into the pair. The input vector is the beam-forming vector of the UE. The input layer is constructed accordingly. In the hidden layers, the neurons are fully connected to all activations in the previous layer. In addition to them, Softmax layer, normalization layers and dropout layers are added. Softmax layer is located in the end of fully connected layer. Softmax layer is used for multiclassification. In the output layer, the output of Softmax layer corresponds to an estimate of the optimal beam allocation. The neural network model is trained by history data about the location of UEs and the true optimal beam pairs. The training data set is generated in the following steps: Step 1: generation of the received signals in terms of UE locations and signal strength; Step 2: collection of the true optimal beam pairs by searching exhaustive method over all beam pairs; Step 3: formulation of the labeled data set from the true optimal beam pairs. Using the trained neural network, we search the beam selection subset, predict the beam pairs with high beam strengths, and find the beam selection subset $\mathrm{D}_k\mathbf{W}_k$ in the test phase. Figure 3.2 illustrates the neural network structure to find the beam selection subset.

As another machine learning method, we consider the K-nearest neighbors (KNN) algorithm where K denotes the nearest neighbor data points

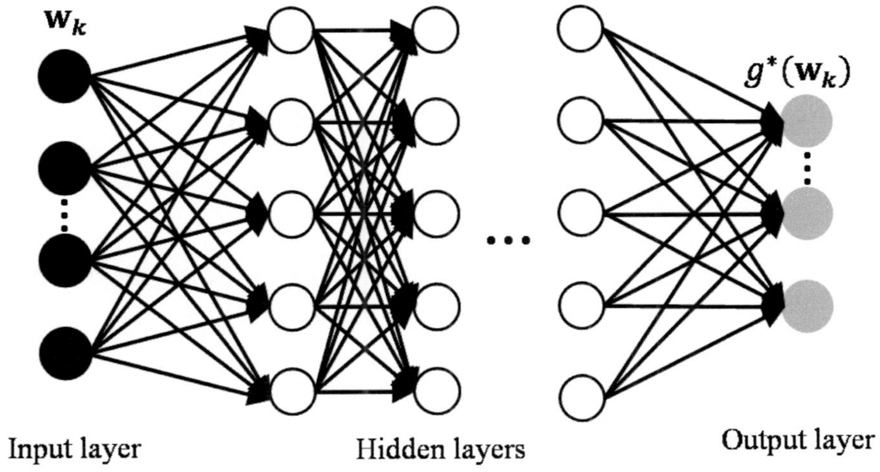

Figure 3.2 The neural network to find the beam selection subset.

(Note, the "K" of KNN is the name of the algorithm and not related to the number of UEs in this chapter). The KNN is simple but powerful to solve both classification and regression problems. It has flexible decision boundaries and is robust to noisy training data. Likewise, we consider the labeled data set as a pair of $(\mathbf{w}_k, g^*(\mathbf{w}_k))$. The value of the target function for a test data set is estimated from the known value of the nearest training data. The input The input $\mathbf{w}_k$ and its nearest neighbour $\mathbf{w}_k{}'$ satisfies

$$\mathbf{w}_k' = \arg\min_{\mathbf{w}_l} \text{dist}\left(\mathbf{w}_k, \mathbf{w}_l\right) \tag{3.11}$$

where $dist$ O is Euclidean distance and $k \neq l$. In the NN classification model, we consider an Euclidean distance and compute the decision boundaries. The decision boundary between any two data points is a straight line. It shows us how input space is divided into classes. The pseudocode of the KNN algorithm can be summarized as follows:

Procedure KNN algorithm

Load the training and test data
Choose the value of κ// *where* κ *is the nearest neighbors and a small positive integer.*
for Each data point in the test data

- Compute the Euclidean distances between the query data points and all training data points
- Store and sort the Euclidean distances
- Choose the first κ points
- Assign a class to the test data point by majority vote

end
return $g^*(\mathbf{w}_k)$.

3.3.3 Optimal beam pair selection using peak finding algorithm

After finding the beam selection subset $\mathrm{D}_k(\mathbf{W}_k)$ by machine learning, eqn (3.10) can be rewritten as follows:

$$g^*(\mathbf{w}_k) = \underset{\mathrm{D}_k(\mathbf{w}_k)}{\operatorname{argmax}} \sum_{k=1}^{K} \log_2\left(1 + \zeta_{k,\mathrm{D}_k(\mathbf{w}_k)}\right), \tag{3.12}$$

where $\zeta_{k,\mathrm{D}_k(\mathbf{w}_k)}$ denotes the beam strength at the k th UE for the beam selection $\mathbf{w}_k$ in the subset $\mathrm{D}_k(\mathbf{W}_k)$. The complexity of the optimization problem eqn (3.12) by an exhaustive search is $O\left(\mathrm{D}_k\left(\mathbf{W}_k\right)\right)$. If we reduce the candidates in the beam selection subset $\mathrm{D}_k\left(\mathbf{W}_k\right)$, the complexity will decrease significantly. Now, we need to select the final beam pair in the subset. Since the size of the subset is not large, the simple peak finding algorithm would be more fit and guarantee the optimal beam pair $\left(\mathbf{w}_k, g^*\left(\mathbf{w}_k\right)\right)$. One of the simple peak finding algorithms is the divide-and-conquer method. First, we divide the problem into multiple subproblems that are smaller data sets. Second, we solve the subproblems iteratively and find them until the subproblems are small enough. Lastly, we combine the results to solve the original problem. The pseudocode of the peak finding algorithm can be summarized as follows:

Procedure Peak finding algorithm
Load the beam selection subset $\mathrm{D}_k\left(\mathrm{W}_k\right)$ where k is $[1, \ldots K]$
Define $m = \lfloor K/2 \rfloor$
for Each data point
if $\mathrm{D}_k(m-1) \leq \mathrm{D}_k(m) \geq \mathrm{D}_k(m+1)$
    return *m*
elseif $\mathrm{D}_k(m-1) > \mathrm{D}_k(m)$ *then*
    return find a peak in $\mathrm{D}_k(1, m-1)$
elseif $\mathrm{D}_k(m) < \mathrm{D}_k(m+1)$ *then*
    return find a peak in $\mathrm{D}_k(m+1, K)$
until One element left in the subset
end

3.4 Numerical Analysis

As we describe the system model in Section 3.2, we consider the single-cell downlink model and formulate the beam selection problem based on GoB and PMI of 3GPP 5G NR. Using a beamforming toolbox and a machine learning toolbox of Matlab [15], we evaluate the performance of the proposed two-stage beam selection method and compare it with the conventional exhaustive search method. The key simulation configurations are summarized in Table 3.1.

3.4.1 Simulation results

In the first simulation, we evaluate the effectiveness of the machine learning algorithms as the beam selection subset finder. According to the size of the

Table 3.1 Simulation configuration.

Parameters	**Values**
Number of cells and transmission	Single-cell downlink
Frequency range	FR1 (sub 6 Hz)
Channel model	MIMO scattering channel with 10 scatterers
Transmitted SSBs	8
Synchronization signal block (SSB)	Case B for FR1
Period of the SS burst	20 ms
Subcarrier spacing	15 kHz
Center frequency	2.5 GHz
Transmit/receive antenna size	8 × 8 (T×)/2 × 2 (R×)
SNR	30 dB
Measurement mode for SSB	Only secondary synchronization signals (SSS)
Number of scatterers	10
Total number of beam pairs	16
Number of different receiver locations in training data set/test data set	200 (training)/100 (test)
Neural network structure	1 input layer, 4 hidden layers with 96 nodes and ReLU activation function, 1 output layer with Softmax layer
The value of ^{k} of KNN	1–16
Beam strength metric	Signal strength (such as RSRP and SINR)
Max number of iterations	1000

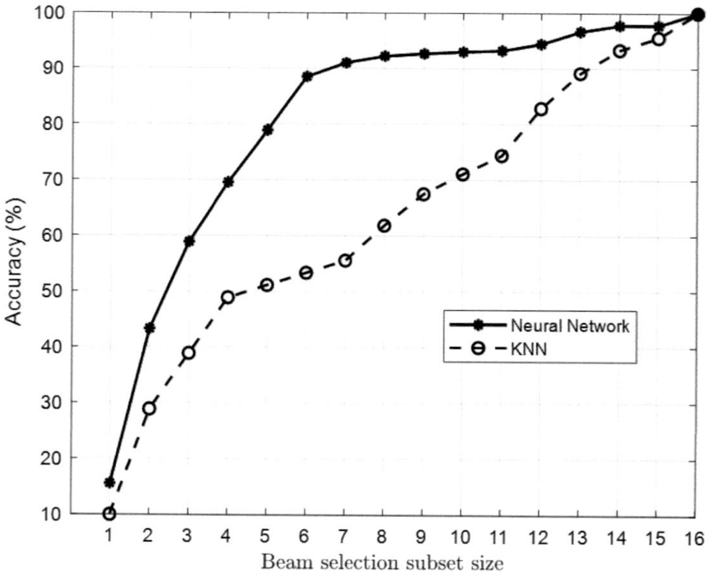

Figure 3.3 Accuracy of the beam pair selection.

beam selection subset, the accuracy of the optimal beam selection is evaluated using the metric of top k accuracy. Figure 3.3 illustrates the accuracy of the beam pair selection methods using deep learning and KNN in terms of the size of the beam selection subset.

As we can observe from Figure 3.3, when the beam selection subset size is 7, the accuracy (%) of the deep learning and KNN is more than 90% and 55%, respectively. This result represents that the trained neural network is useful for finding beam selection subsets and reducing the overhead cost caused by beam sweeping. On the other hand, KNN is not much accurate to find beam selection subsets. When the subset size is 16 (this is identical to the exhaustive search method), the accuracy of machine learning algorithms is the same as the exhaustive search method.

In the second simulation, we evaluate how much the proposed two-stage beam selection method is effective. In order to compare with the exhaustive search method, we compute the average signal strength (dB) of the selected beam pairs when different beam selection methods are used (note: the value of signal strength such as RSRP is typically represented in dBm. However, the purpose of this simulation is to compare their performances. It is represented

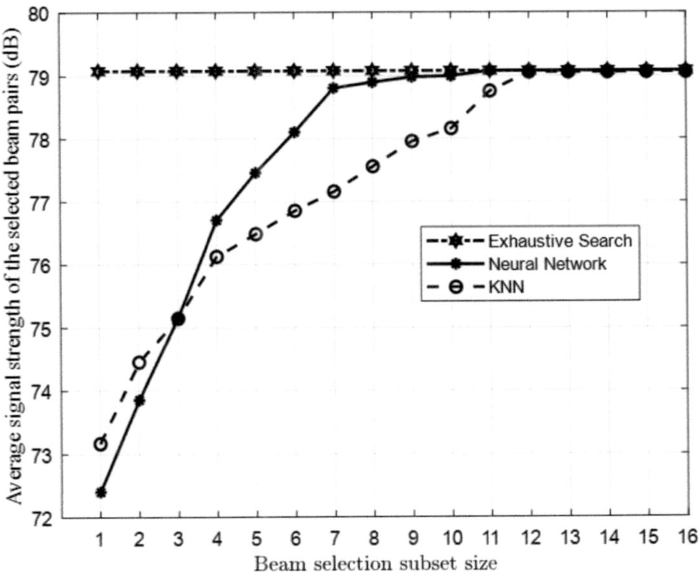

Figure 3.4 The average signal strength of the selected beam pairs by different beam selection methods.

in dB). Figure 3.4 illustrates the average signal strength of the selection beam pairs by different beam selection methods.

As we can observe in Figure 3.4, when the beam select subset size is more than 7, the average signal strength of the selected beam pairs of deep learning is close to the exhaustive search method.

3.4.2 Analysis and lessons learned

From the first simulation, the effectiveness of the neural network in finding the beam selection subset is verified. The neural network model trained by the historical data is useful for finding the optimal beam pair. Under the given simulation configuration, the beam selection subset size should be larger than 7. If we use the neural network, we can find the optimal beam pair in 44% (= 7/16) of the total beam pairs and reduce the overhead cost.

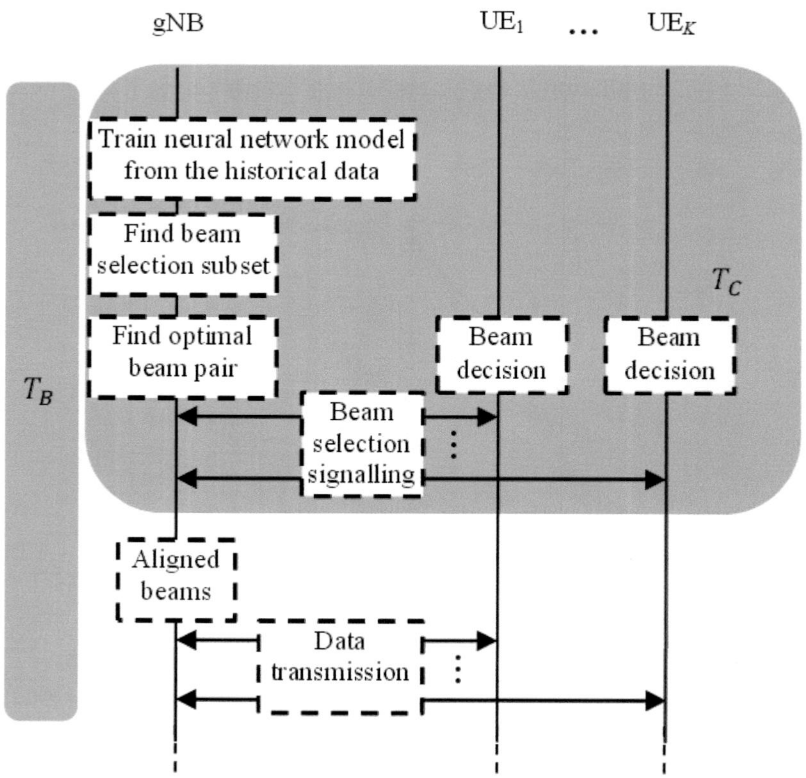

Figure 3.5 Signaling of the proposed two-stage beam selection method with restrictions of coherence times.

From the second simulation, we confirm that the proposed two-stage beam selection method finds the optimal beam pair well from only 44% of beam pair sets. However, another simple machine learning method KNN does not work well. In addition, we need additional computation to train the neural network model.

One important lesson learned from this research work is how training of neural networks affects performance. The MIMO scattering channel model in this simulation is based on a multipath propagation channel depending on time delay, gain, Doppler shift, phase offset, atmospheric loss, and so on. The performance of accuracy varies depending on the channel realizations and different system parameters. The trained neural network does not work well under the different channel conditions and system configurations. In particular, the coherence time should be maintained to have a good level of performance. If the channel is not static, the trained neural network model is not effective, and the accuracy is significantly low. In order to apply the proposed method for a beamforming system, the channel should be treated as constant. Thus, we should consider both a channel coherent time and a beam coherence time and design the beamforming operation. We can design signalling of the proposed two-stage beam selection method as shown in Figure 3.5.

As we can observe in Figure 3.5, the channel coherence time should be maintained during the proposed beam selection method and the data between gNB and UEs are stably transmitted during the beam coherence time.

3.5 Conclusion and Summary

In this chapter, we proposed a two-stage beam selection method consisting of machine learning and peak finding algorithm to reduce the beam sweeping overhead. The exhaustive search method of the 3GPP 5G NR causes a long beam search time, a high overhead cost, and a high energy consumption. The proposed beam selection method predicts the beam selection subset by machine learning and then finds the optimal beam pair from only 44% of the search set. Thus, the UEs do not need to report all PMI information regularly and the beam sweeping overhead can be reduced. By numerical analysis, we showed that the trained neural network model is useful for finding the beam selection subset and the proposed two-stage beam selection method finds the optimal beam pair well. However, the accuracy of predicting the beam selection subset varies in terms of coherent times and system parameters. Thus, future works are to investigate online learning methods of neural networks to satisfy the restriction of coherence times, as well as design refined signaling.

Acknowledgments

This work has been part of the 6G-XR project, which has received funding from the Smart Networks and Services Joint Undertaking (SNS JU) under the European Union's Horizon Europe research and innovation program under Grant Agreement No 101096838.

References

[1] 3GPP TR 38.802 - 6.1.6.1 Beam Management.

[2] W. Zirwas, M. B. Amin, and M. Sternad, "Coded CSI reference signals for 5G — Exploiting sparsity of FDD massive MIMO radio channels," *in Proc. IEEE WSA*, Mar. 2016.

[3] 3GPP, "NR; physical layer procedures - Rel. 15," TS 38.214. Dec. 2018.

[4] 3GPP, "TR 36.897: Technical Specification Group Radio Access Network; Study on elevation beamforming / Full-Dimension (FD) Multiple Input Multiple Output (MIMO) for LTE (Release 13)," 3GPP, Tech. Rep. V13.0.0, 2015

[5] E. Dahlman, S. Parkvall, and J. Sklöld, *5G NR: the Next Generation Wireless Access Technology*, Academic Press, 2018.

[6] R. Pal, K. V. Srinivas, and A. K. Chaitanya, "A beam selection algorithm for millimeter-wave multi-user mimo systems," *IEEE Communications Letters*, vol. 22, no. 4, pp. 852–855, April 2018.

[7] Y. Niu, Z. Feng, Y. Li, Z. Zhong, and D. Wu, "Low complexity and near optimal beam selection for millimeter wave mimo systems," *in 2017 13th International Wireless Communications and Mobile Computing Conference (IWCMC)*, June 2017, pp. 634–639.

[8] J. Wang, "Beam codebook based beamforming protocol for multi-Gbps millimeter-wave WPAN systems," *IEEE J. Sel. Areas Commun.*, vol. 27, no. 8, pp. 1360–1399, Oct. 2009.

[9] S. Hur, T. Kim, D. J. Love, J. V. Krogmeier, T. A. Thomas and A. Ghosh, "Millimeter wave beamforming for wireless backhaul and access in small cell networks," *IEEE Trans. Commun.*, vol. 61, no. 10, pp. 4391–4403, Oct. 2013.

[10] W.Wu,D. Liu, Z. Li, X. Hou, and M. Liu, "Two-stage 3Dcodebook design and beam training for millimeter-wave massive MIMO systems," *in Proc. IEEE 85th Veh. Technol. Conf.*, 2017, pp. 1–7.

[11] Y. Heng and J.G. Andrews, "Machine Learning-Assisted Beam Alignment for mmWave Systems," *in Proc. 2019 IEEE Global Communications Conference (GLOBECOM)*, 2019, pp. 1-6, doi:10.1109/GLOBECOM38437.2019.9013296.

[12] A. Klautau, P. Batista, N. González-Prelcic, Y. Wang, and R.W. Heath, "5G MIMO Data for Machine Learning: Application to Beam-Selection Using Deep Learning," *in Proc. 2018 Information Theory and Applications Workshop (ITA)*, 2018, pp. 1-9, doi:10.1109/ITA.2018.8503086.

[13] V. Va, J. Choi, and R. W. Heath, "The impact of beamwidth on temporal channel variation in vehicular channels and its implications," *IEEE Transactions on Vehicular Technology*, vol. 66, no. 6, pp. 5014-5029, June 2017, doi:10.1109/TVT.2016.2622164.

[14] T. M. Pham, R. Farrell and L. N. Tran, "On Estimating Maximum Sum Rate of MIMO Systems with Successive Zero-Forcing Dirty Paper Coding and Per-antenna Power Constraint," *in Proc. 2019 IEEE 30th Annual International Symposium on Personal, Indoor and Mobile Radio Communications (PIMRC),* Istanbul, Turkey, 2019, pp. 1-6, doi:10.1109/PIMRC.2019.8904416.

[15] https://mathworks.com/

4

A Smart Trouble Ticket Management Approach for Mobile Networks Using Unsupervised Learning

Thaina Saraiva[1, 2], David Duarte[1, 2], Luis Mata[1, 2, 3, 4], and Pedro Vieira[2, 4]

[1]CELFINET (a Cyient Company), Lisbon, Portugal
[2]Instituto de Telecomunicações (IT), Lisbon, Portugal
[3]Instituto Superior Técnico (IST), Universidade de Lisboa, Lisbon, Portugal
[4]Instituto Superior de Engenharia de Lisboa (ISEL), Lisbon, Portugal

Abstract

The increasing complexity of mobile networks as they evolve towards 6th Generation (6G), coupled with the importance of providing fast responses to performance degradation issues, requires the adoption of cost-effective Trouble-Ticket (TT) management systems by Mobile Network Operators (MNOs). These systems are pivotal in ensuring compliance with Service Level Agreements (SLAs) and maintaining Quality of Service (QoS) standards. In this context, the introduction of a new Smart Operations (SO) framework referred to as Smart Trouble Ticket Management (STTM), promotes the integration of Artificial Intelligence (AI)/Machine Learning (ML) to leverage performance data, aiming at early performance degradation detection and an efficient handling of TTs. This paper presents a novel methodology for STTM that relies on a ML clustering algorithm to identify performance degradation instances when cells transition to lower-ranked clusters. The benefits of this dynamic approach are two-fold. Firstly, it offers an automated and adaptable process capable of accommodating diverse Key Performance Indicators (KPIs). Secondly, it enhances the reliability of Trouble-Ticket (TT) opening and closure compared to conventional approaches, particularly those

DOI: 10.1201/9788743801108-5

reliant on fixed threshold values. The proposed methodology, tested on live network data, achieved an improvement of 5.53% in TT Opening Accuracy Rate (OAR), and reduced the TT Repeat Call Rate (RCR) by 17.57%.

Keywords: Mobile Networks, Smart Operations, Smart Trouble-Ticket Management, Performance Management, AI/ML

4.1 Introduction

Considering the increase in network complexity and stringent Quality of Experience (QoE) requirements on the road to 6th Generation (6G), Mobile Network Operators (MNOs) are expected to adopt Smart Operations (SO) as a new operative model. SOs encompasses various applications that leverage network data and Artificial Intelligence (AI)/Machine Learning (ML) to enhance operational efficiency while promoting sustainable resource utilization. A recent example of a research initiative on SOs is the CANOPY project [1].

In the traditional reactive mode approach, troubleshooting only starts after a new service fault is reported and processed, leading to longer Mean Time to Repair (MTTR), lower service availability, and increased customer complaints. Therefore, the challenge for MNOs lies in enhancing Network Operations Centers (NOCs) automation to swiftly detect, diagnose and recover from performance degradation incidents while upholding trust and reliability, which are vital in sustaining the economic impact of mobile networks. A crucial application for SOs is addressing performance degradation issues, aiming to identify cells falling below Service Level Agreement (SLA) requirements. This triggers the opening of a Trouble-Ticket (TT), which typically involves analyzing multiple data sources, including Performance Management (PM) data and customer complaints. Despite some previous attempts to integrate data sources and automate manual repetitive tasks, the traditional automatic performance degradation detection methods rely on fixed thresholds setting to opening TTs, the definition of acceptable variance intervals for KPIs, or a combination thereof, as exemplified in [7]. However, these approaches lack adaptability across diverse network elements with distinct performance profiles. Moreover, it proves challenging to manage and relies solely on empirical knowledge. Hence, an adaptive STTM dynamic approach becomes essential.

In this paper, a new STTM approach is presented, which dynamically adjusts the performance thresholds and variation intervals used for handling

TTs, thus enabling NOCs to provide support for multiple services and multi-tenancy. The proposed methodology offers versatility across technologies, combining the MNO's empirical knowledge with statistical learning based on PM data. It adopts a comprehensive two-step process, starting with a setup stage that utilizes historical PM data to identify cell performance patterns. Subsequently, an operational stage extracts updated PM indicators from the Operations Support System (OSS), serving as input for the decision algorithm to dynamically open or close TTs in response to performance changes. In summary, the main contributions of this paper are:

- Introduction of a new, agnostic, flexible, and dynamic classification me-thodology to facilitate the decision-making process for opening or closing new TTs in response to performance degradation events.
- Evaluation of the proposed methodology through the utilization of live network data, employing an unsupervised ML clustering algorithm.

The paper is organized as follows: Following the introduction, Section II explores related work and introduces the proposed STTM methodology. Section III presents the experimental study, using live network data to evaluate the proposed methodology, while comparing it to a traditional fixed threshold approach. In Section IV, conclusions are drawn, and future work is outlined.

4.2 Research Methodology

This section provides an overview of related work and describes the STTM framework proposed in this paper.

4.2.1 Related Work

Trouble Ticket Management (TTM) systems are designed to manage the creation, updating, and resolution of incident tickets related to abnormal network events, such as the case of handling performance degradation issues in mobile networks.

Considering the recent success of AI/ML techniques to leverage network data to increase operational efficiency, numerous research papers and practical implementations have explored the use of statistical learning to detect performance degradation and anomalies in mobile networks, as in [12, 6]. In light of the challenges associated with labelling target variables for performance degradation scenarios, many studies within the field have turned to unsupervised learning methods to address this issue. An example is provided in [11], where an unsupervised learning approach for anomaly detection related to mobility issues in 4G networks is introduced. This framework, which combines Principal Component Analysis (PCA) with Hierarchical Density-Based Spatial Clustering (HDBSCAN) to identify performance degradation in cells, demonstrated that clustering algorithms are highly efficient in understanding and detecting anomalous mobility behaviour in mobile networks. In [10], an automated decision tree approach is employed to streamline operator troubleshooting, with the primary goal of reducing false positive TT generation triggered by a cell's KPI surpassing predefined thresholds. This method effectively distinguishes between tickets associated with known causes, automatically pinpointed through field measurements, and those needing expert intervention due to external factors. In [13], the focus shifts to unsupervised clustering techniques, encompassing K-means and hierarchical clustering, for anomaly detection. These methods are effective in identifying unusual surges in traffic demand at specific locations and times within the network.

Recent advancements, as in [5], revolve around enhancing autonomous anomaly detection within Self-Organizing Networks (SON) through user-centric methodologies and ML techniques. This research contributes to improving network self-healing mechanisms and bolstering energy efficiency by selectively powering down malfunctioning nodes based on user experience ratings. A variety of ML approaches are explored, aiming for scalability and optimal effectiveness. In [2], a novel approach utilizing autoencoders, a class of deep neural networks, is proposed for detecting anomalies in 4G performance. This study investigates the influence of various autoencoder parameters and recommends suitable architectures to achieve robust anomaly detection. Furthermore, [3] delves into the analysis of network performance data at both daily and hourly intervals, shedding light on the advantages of employing supervised ML techniques relying on random forest and support vector machines for anomaly detection in mobile networks.

Numerous AI/ML approaches have been explored for handling performance degradation in mobile networks. However, they often struggle to effectively combine statistical learning with Mobile Network Operators (MNOs)' empirical knowledge, resulting in highly specific, and less interpretable outcomes that can deter adoption in real use cases. The proposed framework bridges this gap by offering a tailored solution that integrates automation while leveraging on domain empirical knowledge. Its versatility makes it suitable for diverse problem types, and it places a strong emphasis on delivering clear, explainable results.

4.2.2 Problem Description

The NOC employs a standardized procedure for accepting site installations or network parameter configuration changes, involving continuous monitoring to detect potential cell performance degradation. Traditionally, this monitoring relies on threshold-based methods, where Key Performance Indicators (KPIs) are continuously tracked within specific intervals, guided by predefined SLA targets. For example, the MNO may aim for 99% call retainability per day. When a monitored KPI falls below its threshold, an alarm is triggered and a trouble ticket (TT) is opened to indicate inadequate cell performance. Experts then analyze and resolve the issue, as described in [10]. However, this process can sometimes lead to multiple consecutive TT re-openings, introducing complexity and increasing manual analysis costs. Therefore, optimizing performance degradation detection and minimizing repeated TT re-openings are essential for enhancing efficiency and reducing expenses.

4.2.3 Proposed STTM Framework

The proposed STTM framework aims at detecting cell-level network performance degradation by daily monitoring of KPIs, primarily using PM or CM data. It combines two complementary approaches: a threshold-based and a cluster-based. The first approach focuses on short-term KPI variation, comparing average KPI values at regular intervals (*e.g.,* daily) vis-à-vis predefined absolute thresholds. Conversely, the second approach identifies long-term performance degradation by considering average performance in comparison to similar cells.

The methodology execution starts with data acquisition from PM and Configuration Management (CM) sources. The preprocessing phase involved removing outliers, and feature engineering was conducted to include additional variables. These variables, such as low traffic and cell unavailability, are instances where pseudo anomalies can occur. Subsequently, it proceeds through two stages: the setup and the operational. During the setup stage, historical data is employed to understand network cell behaviour patterns. Over time (*e.g.,* weekly), cells are classified into distinct clusters, serving as a reference until a new setup stage is executed (*e.g.,* after a network deployment). After the initial setup stage, the operational mode is initiated, involving the upload of new data and the analysis of cell performance changes. This phase aims to detect cell cluster transitions, which can trigger actions such as opening, updating, or closing TTs. In Figure 4.1, the setup stage and

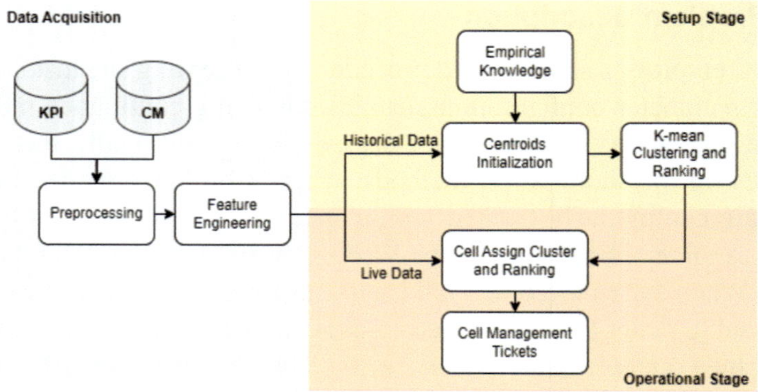

Figure 4.1 STTM System Architecture.

operational stages are depicted, highlighting the use of clustering techniques to identify cell behaviour patterns and group objects based on their similarity.

Here, the K-means algorithm, often utilizing Euclidean distance as the metric [8], is commonly applied. However, sensitivity to initial centroid placement can affect results, as noted in [14]. To address this challenge, a predefined number of initial clusters is considered. Empirical knowledge regarding reference performance levels is also incorporated to fine-tune the clustering parameters. The optimal number of clusters is ultimately determined by evaluating their distinctiveness using Euclidean distance and comparing them to historical data.

During the operational stage, the algorithm continuously monitors and evaluates daily data collected from the live network. The TT management system actively handles cell performance by opening, updating, and closing tickets as needed. The process of initiating a TT starts by checking the threshold value. If the threshold is not breached, the clustering method is applied, utilizing data from the past week to assign cells to clusters. This cluster can either be an existing one or a new cluster. However, if the cell is assigned to a cluster with worse performance, a TT is opened based on the reason for the cluster transition. The TT is subsequently updated when there is a change in the degradation reason, which can occur either due to threshold breaching or cluster transitioning towards a worse performance cluster. The TT is modified to reflect the updated degradation reason. Finally, the TT is closed when the considered KPI returns to a cluster of better performance and no longer violates the threshold value. The STTM methodology generates three types of TTs. The first is "Type 1," encompassing TTs that only violates the threshold criteria. "Type 2" includes TTs not detected by the threshold but successfully identified through the clustering approach. "Type 3" covers hybrid TTs with updates for both threshold and clustering criteria during their active open period.

From the standpoint of the NOC, the opening of a new TT draws a parallel with the receipt of a customer issue complaint within a call centre. However, within the framework of a mobile network, a single TT can encompass the complaints of a multitude of affected customers, thereby underscoring the paramount significance of efficient TT management. To optimize resource allocation efficiency and proactively discern and rectify persistent issues, a process must be implemented by the NOC to monitor the proportion of issues requiring subsequent interactions or further engagements following the initial contact. The primary objective in this context is to enhance the resolution of customer issues upon their initial contact. In this paper, the Repeat Call Rate (RCR), a metric commonly used in call-centre environments, was employed to evaluate the proposed STTM methodology over a 30-day period, as defined in [4].

$$RCR = \frac{R}{N}. \tag{4.1}$$

Where N is the total number of TTs processed over a consecutive 30-day period, and R is denotes the number of repeated TT (*i.e.,* focusing on the same issue) over that period. Formally, R can be defined as:

$$R = \sum_{j=0}^{N-1} R_j, \tag{4.2}$$

where $j \in \{0, 1, \dots, N-1\}$ denotes each individual TT index and R_j is computed as follows:

$$R_j = \left\{ \begin{array}{ll} 0 & \text{First TT opened after KPI degradation} \\ 1 & \text{Other TTs opened in the next 30 days} \end{array} \right\}.$$

Considering the NOC context in monitoring mobile network service performance, an indicative maximum holding time of 30 days is proposed for a new TT, which can be adjusted if necessary. Hence, if a problem occurs within this time frame, it is assumed to be related to the initial problem. However, if the interval exceeds 30 days, it is considered a separate and distinct issue. This approach effectively tracks and addresses recurring problems while ensuring unrelated issues are appropriately managed. Therefore, a customer call is considered equivalent to a new TT opened due to KPIs degradation detected by the ticket management system.

Moreover, if a TT remains open for a period longer than 30 days, then the algorithm initiates the creation of a "virtual" new TT, even if the original one has not yet been closed. This approach is based on the principle that there is a reasonable time frame within which a TT should ideally be resolved, here set to 30 days. However, this duration can be adjusted to align with the specific operational requirements of each NOC. By introducing the concept of the "virtual" TT, the proposed algorithm ensures that any existing customer complaint or performance issue is carried forward into the subsequent 30-day window. This mechanism effectively treats new tickets generated during this period as repetitions rather than entirely new events. This approach enhances the ability to track and address recurring problems systematically while distinguishing them from unrelated issues.

In summary, the STTM approach offers advantages over the threshold-based method, including enhanced performance analysis, improved incident management, and customisable time frames for "virtual" TTs to meet NOC's requirements. In Figure 4.2 a comparison between both approaches is depicted.

In the first 30-day period, the STTM approach introduced just one TT, which remained unresolved throughout the entire period, resulting in an RCR of 0% ($N = 1$, $R = 0$). Conversely, the traditional threshold-based method

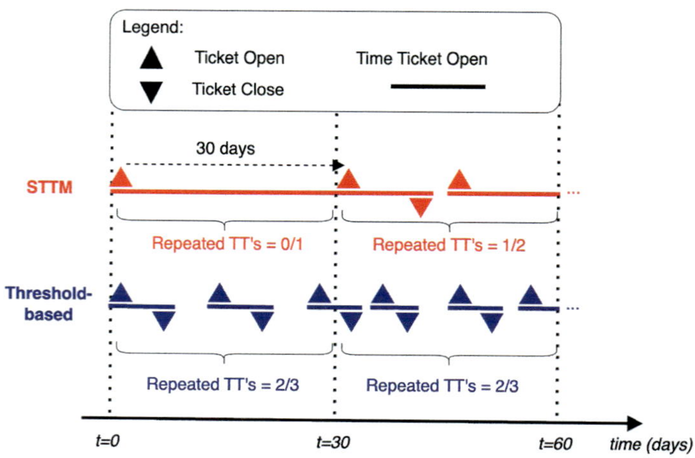

Figure 4.2 Proposed STTM methodology highlighting the comparison with the threshold-based approach to handle trouble tickets.

generated three separate TTs, with the first one representing a genuine issue and the subsequent two being redundant repetitions. This led to a RCR of 67% ($N = 3$, $R = 2$). This discrepancy underscores that while the traditional threshold-based approach can exhibit significant volatility due to frequent threshold breaches, the STTM approach employs a time-averaging technique for KPIs, which reduces the occurrence of repeated incidents.

4.3 Experimental Study

This section presents the dataset used for the evaluation of the proposed STTM methodology. Moreover, a comparison with the traditional threshold-based approach is presented using five representative KPIs and the results are discussed.

4.3.1 Dataset Description

The analysis was based on real data, including 21 4G sites and 161 cells, as described in Table 4.1. Data extraction was conducted on a daily basis, spanning a 90-day duration.

The dataset encompassed the monitoring of five KPIs for the access network, as detailed in [9]:

(1) **Average CQI**: evaluates the quality of the downlink radio channel.
(2) **Call Setup Success Rate (CSSR)**: the percentage of successfully established calls.
(3) **Downlink Resource Block Utilization**: the utilization of downlink resource blocks.
(4) **Service Drop Rate (SDR)**: is the rate at which service drops occur during daily operations.
(5) **User Downlink Average Throughput**: average data transfer rate for users in the downlink direction.

Table 4.1 Frequency bands from the 161 cells.

Frequency [MHz]	Bandwidth [MHz]	Number of Cells
700	10	12
800	10	50
900	10	4
1800	20	65
2100	10	21
2600	20	9

4.3.2 Trouble Tickets Assigmnment

During the setup stage (7 days of data from March 1st to March 7th), historical data was utilized to generate features for the K-means algorithm. This algorithm assigns reference cluster centroids that represent groups of cells with similar performance. In the operational stage, cells are assigned to clusters based on their shortest Euclidean distance. Subsequently, a performance monitoring phase is initiated through ticket management (opening, updating, and closing of TTs). Figure 4.3 illustrates the performance evolution of the SDR KPI, over a 20-day period.

The acceptable threshold considered for the SDR was defined as lower than or equal to 1%. Within the initial highlighted section, there is a "Type 1" TT that occurred on the 10th April. This TT was opened because the KPI value exceeded the established threshold. On the following day, the TT was closed as the KPI value dropped below the threshold. The second highlighted area on the chart corresponds to a “Type 3” TT. In the setup stage, two clusters were defined: “Cluster 0” and “Cluster 1”. The first cluster represents cells with higher performance, while the second represents cells with lower performance. Before the 14th April, the cell was assigned to “Cluster 0”. However, from this day onwards, the cell’s performance started to deteriorate, leading to the opening of a TT based on clustering analysis, where the cell transitioned to “Cluster 1”. However, on the 19th April, the KPI once again exceeded the threshold. Since there was already an open TT during this period, the TT reason was updated to reflect the threshold violation. Finally, on the 26th April, considering the cell’s performance over the past week, it was assigned to a higher performance cluster and the TT was closed.

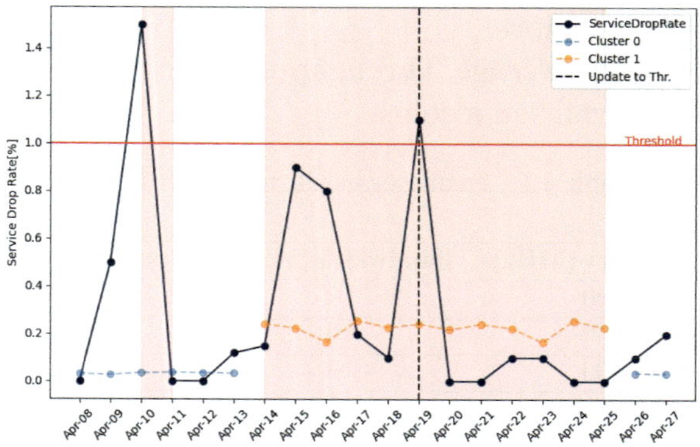

Figure 4.3 Service Drop Rate KPI monitoring.

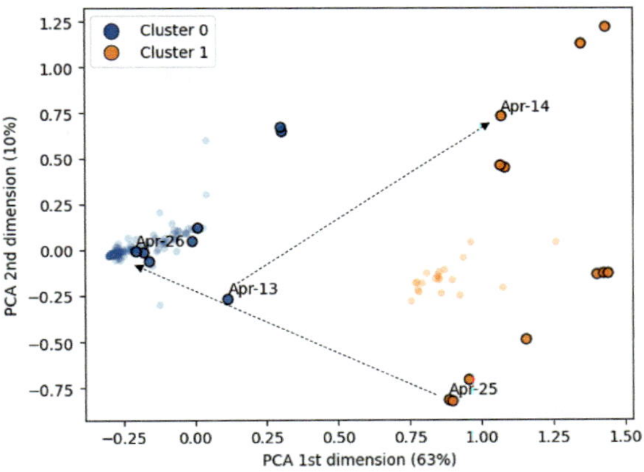

Figure 4.4 Example of PCA to analyse a cell daily cluster assignment.

To enhance the interpretability of the previous example, which examines the evolution of the SDR KPI, a PCA was employed to delve into the daily cluster transitions. This analysis provides a more comprehensive understanding of the underlying data patterns, by reducing feature dimensionality. In Figure 4.4, the daily progression using a PCA is presented, which projects the data into two dimensions, capturing 73% of the variance (respectively, 63% in the 1st and 10% in the 2nd). This visualization highlights the daily cluster allocation of a single cell, initially assigned to cluster 0, and its subsequent transitions.

This analysis focuses on two consecutive periods ($13 - 14^{th}$ April, and $25{-}26^{th}$ April). As observed, at the end of the first period the cell transitioned to cluster 1, and then went back to its initial cluster, after the second period. In conclusion, PCA complements the STTM approach by offering improved clarity regarding the present performance status and its dynamic evolution over time.

4.3.3 Results' Analysis

A comparison of the three types of TT identified using the STTM methodology is presented in Table 4.2, comprising 253 TT, that were identified across 78 cells spanning 20 distinct sites (i.e. geographical locations). Among the 253 TTs, 14 were exclusively detected by the clustering method, and would not have been identified by the threshold-based approach, indicating an Opening Accuracy Rate (OAR) of 5.53%. Additionally, "Average CQI" and "CSSR" had the most TTs, with 127 and 69 TTs, respectively.

Table 4.2 Number of opened tickets by type and KPI.

KPI	Type 1 (Threshold)	Type 2 (Clustering)	Type 3 (Hybrid)
(1) Average CQI	76	0	51
(2) Call Setup SR	36	0	33
(3) DL Resource Block UR	18	12	13
(4) Service Drop Rate	3	1	7
(5) User DL Avg. Thr.	0	1	2
Total	133	14	106
Grand Total		**253**	

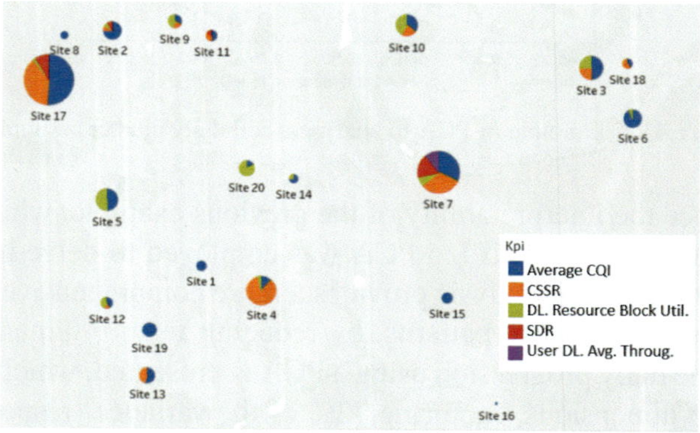

Figure 4.5 Aggregated TT by KPI across the analysed area.

In Figure 4.5, each pie segment's size corresponds to the total count of generated TTs at all 20 sites, considering the five addressed KPIs.

Notably, 'Site 17' stands out with the highest TT count, indicating issues in four KPIs. Specifically, the 'Average CQI' KPI displayed degradation across most sites, while 'User DL Average Throughput' TTs were specific to 'Site 7'.

4.3.4 STTM vs Threshold-based Approach

A thorough comparison of the STTM methodology and the traditional threshold-based approach was conducted using real 4G network data, focusing on the RCR metric, as described in Section 4.2. Figure 4.6 summarizes their comparative performance in managing trouble tickets (TTs) over 90 days, specifically for the CSSR KPI in five cells (A, B, C, D, and E).

During the 90-day observation period, all cells in question experienced the opening of TTs under both compared approaches. Except for "Cell

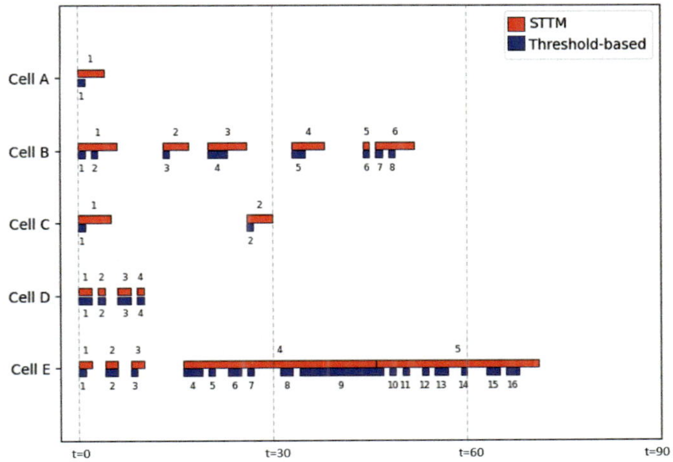

Figure 4.6 Comparison between Threshold-based and STTM approach for CSSR KPI.

A," multiple tickets were generated in all cells, resulting in a non-zero RCR. Notably, the STTM approach reduced the RCR. For example, in the first 30-day period, "Cell E" had 6 repeated TTs with the threshold-based approach and only 3 with the STTM approach, resulting in RCR values of 85% (6/7) and 75% (3/4), respectively. In the second 30-day period, "Cell E" had no repeated TTs with the STTM method, whereas the threshold-based approach had 6 repetitions. Furthermore, it is important to highlight that, as outlined in Section 4.2, a new "virtual" TT with reference 5 was generated for "Cell E" during period 2. This was based on the premise that the total duration of a TT should not exceed 30 days, in accordance with a reasonable resolution window criterion. Finally, in period 3, "Cell E" had only one repeated TT with the threshold-based method. Table 4.3 provides a comparison of the two methods based on the average RCR for the analysed KPIs.

Table 4.3 Average RCR by KPI and Method.

	RCR [%]		
KPI	**THR-based**	**STTM**	**Change (%)**
(1) Average CQI	35.16	15.67	19.49
(2) Call Setup SR	39.21	27.44	11.77
(3) DL Resource Block UR	52.72	27.46	25.26
(4) Service Drop Rate	20.83	17.85	2.98
(5) User DL Avg. Thr.	53.33	25.00	28.33
Total	**40.25**	**22.68**	**17.57**

The results show that the threshold-based approach yielded a higher RCR compared to the STTM. The overall average was 40.25% for the threshold-based approach and 22.68% for the STTM, resulting in a reduction of 17.57% of the RCR.

4.4 Conclusions and Future Work

This paper presents an automated methodology for managing performance TT in radio-access mobile networks. It complements the traditional fixed threshold-based method, by incorporating unsupervised learning techniques and considering network-specific characteristics. The methodology utilizes the K-means algorithm for detecting cell patterns and employs live monitoring to track cluster transitions and identify cell degradation.

The STTM methodology proved to be a valuable approach in performance analysis by considering patterns and clusters, rather than solely relying on individual thresholds. This approach enabled a more comprehensive understanding of performance degradation, enabling the more consistent identification of cells with degradation, and facilitating the exploration of underlying causes. The STTM approach not only reduced RCR by 17.57% compared to the traditional threshold method, but also detected an additional 5.53% of tickets that would have been forgotten by the threshold-based approach. These findings highlight the effectiveness and benefits of employing the STTM methodology in performance monitoring and management.

In conclusion, using trouble-ticketing for KPIs enables MNO to proactively manage network performance, detect anomalies, and identify areas for improvement to ensure optimal service quality. Future work will focus on integrating energy consumption data and on improving the degradation detection framework with enhanced root-cause analysis. With the identification of underlying causes, a comprehensive understanding of performance issues will foster more effective troubleshooting by ascertaining root causes and implementing targeted problem-solving strategies.

Acknowledgements

The authors would like to acknowledge the importance of EUREKA Clusters project AI Call 2021 (AI2021-061 CANOPY), supported by the PORTUGAL 2020 Partnership Agreement (SI I&DT Empresarial - Copromoção N° 181417). This work has also received funding from UIDB/00760/2020. Moreover, a heartfelt acknowledgement is due to the

Instituto de Teleco-municações (IT) and CELFINET (a Cyient Company), who made this work possible.

References

[1] CANOPY: Cognitive and Automated Network Operations for Present and Beyond, 2023. Accessed: 2023-04-13.

[2] Md Rakibul Ahasan, Mirza Sanita Haque, Mohammad Rubbyat Akram, Mohammed Fahim Momen, and Md Golam Rabiul Alam. Deep learning autoencoder based anomaly detection model on 4G network performance data. *2022 IEEE World AI IoT Congress (AIIoT)*, pages 232–237, 2022. doi: 10.1109/AIIoT54504.2022.9817338.

[3] Md Rakibul Ahasan, Mirza Sanita Haque, and Md Golam Rabiul Alam. Supervised learning based mobile network anomaly detection from KPI data. *2022 IEEE Region 10 Symposium (TENSYMP)*, pages 1–6, 2022. doi: 10.1109/TENSYMP54529.2022.9864371.

[4] Elma Avdagié-Golub, Nermin Goran, Belma Memić, Muhamed Begović, and Bakir Karahodža. A new evaluation method for call rating in contact center. *2022 30th Telecommunications Forum (TELFOR)*. pages 1–4, 2022. doi:10.1109/TELFOR56187.2022.9983674.

[5] S Caleb and S. John Justin Thangaraj. Anomaly detection in self-organizing mobile networks motivated by quality of experience. *2023 Fifth International Conference on Electrical, Computer and Communication Technologies (ICECCT)*, pages 1–6, 2023. 10.1109/ICECCT56650.2023.10179726.

[6] Yeon-Jea Cho, Young-Seok Kim, Sunghyun Kim, Dongkyu Sim, Doyoung Kwak, and Jongsik Lee. Ai-enabled wireless kpi monitoring and diagnosis system for 5G cellular networks. *2019 International Conference on Information and Communication Technology Convergence (ICTC)*, pages 899–901, 2019. 10.1109/ICTC46691.2019.8939988.

[7] Madalena Cilínio, David Duarte, Pedro Vieira, Maria Paula Queluz, and AntÃşnio Rodrigues. Root cause analysis of low throughput situations using boosting algorithms and the treeshap analysis. *2022 IEEE 95th Vehicular Technology Conference: (VTC2022-Spring)*, pages 1–5, 2022. 10.1109/VTC2022-Spring54318.2022.9860734.

[8] T. Cover and P. Hart. Nearest neighbor pattern classification. *IEEE Transactions on Information Theory*, 13(1):21–27, 1967. 10.1109/TIT.1967.1053964.

[9] Huawei. Accessibility KPIs. *Service Integrity*, pages 1–4, Int. Rep., 2020.

[10] Anand Padmanabha Iyer, Li Erran Li, and Ion Stoica. Automating diagnosis of cellular radio access network problems. *Proceedings of the 23rd Annual International Conference on Mobile Computing and Networking*, page 79–87, 2017. doi: 10.1145/3117811.3117813.

[11] Jessica Moysen, Furqan Ahmed, Mario García-Lozano, and Jarno NiÃńmela. Unsupervised learning for detection of mobility related anomalies in commercial LTE networks. *2020 European Conference on Networks and Communications (EuCNC)*, pages 111–115, 2020. 10.1109/EuCNC48522.2020.9200970.

[12] David Mulvey, Chuan Heng Foh, Muhammad Ali Imran, and Rahim Tafazolli. Cell fault management using ML techniques. *IEEE Access*, 7:124514–124539, 2019. doi: 10.1109/ACCESS.2019.2938410.

[13] Md Salik Parwez, Danda B. Rawat, and Moses Garuba. Big data analytics for user-activity analysis and user-anomaly detection in mobile wireless network. *IEEE Transactions on Industrial Informatics*, 13(4):2058–2065, 2017. 10.1109/TII.2017.2650206.

[14] Lulu Wen, Kaile Zhou, and Shanlin Yang. A shape-based clustering method for pattern recognition of residential electricity consumption. *Journal of Cleaner Production*, 212:475–488, 2019. 10.1016/j.jclepro. 2018.12.067.

5

A Q-learning Algorithm-based Frequency Resource Allocation Scheme for User Links in High Throughput Satellite Communication

Go Otsuru[1,2], Yukitoshi Sanada[2], Hiroyuki Tsuji[1], and Amane Miura[1]

[1]National Institute of Information and Communications Technology, Japan
[2]Keio University, Japan
E-mail: otsuru@nict.go.jp; sanada@elec.keio.ac.jp; tsuji@nict.go.jp; amane@nict.go.jp

Abstract

High-throughput satellites (HTS) with frequency flexibilities have attracted increasing attention due to their high-capacity and high-speed communication, which enables a lower cost per bps for end users. In this chapter, we model a beam resource allocation problem as a beam-series finite Markov decision process and propose a frequency allocation scheme for beams in HTS communication systems using a Q-learning algorithm. Under the constraint of the limited bandwidth allocated to the beam, the frequency resource allocations to user equipment (UE) are conducted using a dynamic programming algorithm, which is known to provide an efficient solver for 0-1 knapsack problems in order to obtain an optimal solution according to the policies of a network operation center. Here, the state of the beam is determined from the frequency allocation in a previous beam, the action of the beam is defined as an allocated bandwidth, and the rewards to update the Q-table are calculated from the link allocations. The proposed scheme was evaluated through computer simulation and compared with a full search scheme and an immediate reward maximization scheme. The proposed allocation scheme outperforms the full search scheme in terms

DOI: 10.1201/9788743801108-6

of computational complexity. In addition, simulation results show that the proposed scheme outperforms the immediate reward maximization scheme. The system throughput of the proposed allocation scheme reached 96.1% of the full search, and that of the immediate reward maximization scheme reached only 80.8% when a policy that maximizes the system throughput was adopted. In addition, with the proposed allocation scheme, the average number of the allocated UE was 96.1% of the full search, and that of the immediate reward maximization scheme reached only 76.7% when a policy that maximizes fairness among the UE was adopted.

Keywords: High-throughput satellite, satellite communication, resource allocation, Q-learning, dynamic programming

5.1 Introduction

The traffic demands of satellite communications (SATCOM) are growing continually. In addition, the Satellite Industry Association has reported that the total revenues of the global space economy were 386 billion dollars in 2021, of which satellite services represented 118 billion dollars, ground equipment represented 142 billion dollars, satellite manufacturing represented 6.2 billion dollars, and the launch industry represented 5.7 billion dollars [1]. Increasing affordability supports the growth of this large market; thus, it is essential to reduce the cost per bps for end users, which is expected to improve the capacity of each satellite. High-throughput satellites (HTS) are expected to possess flexible abilities, including variable bandwidth allocated to beams, variable beam shape, 100–1000 beams via multibeam antennas (MBA), the Ka-band for wide bandwidth, and site diversity, through digital processing technologies.

Digital channelizers enable flexible resource allocation in bent-pipe satellites [2–4]. To fully leverage the frequency flexibility of HTSs and improve spectral efficiency (SE), frequency resource allocation schemes based on artificial intelligence (AI) have been investigated [4–7]. For example, a previous study [4] proposed a system analysis model to evaluate the flexibility of a satellite communication system, and the flexibility of a hybrid transponder with a digital channelizer was evaluated. Another study [5] proposed a frequency bandwidth allocation method based on the model predictive control for a satellite communication system. The authors of [6] adopted a deep Q-network in multifrequency time division multiple access systems to solve the frequency allocation problem. In [7], resource management was broken down into a two-stage machine learning approach with a beam-domain resource configuration that included power and gain control, as well as user-domain resource configuration that focused on the bandwidth allocation under the assumption of soft frequency reuse.

Part of the corresponding results has already been presented in [8].

In the current study, we modeled the frequency beam resource allocation problem as a beam-series finite Markov decision process (MDP) rather than a time-series finite MDP because the allocation to a specific beam is dependent on the allocation to the adjacent beam. Here, a Q-learning algorithm is employed to maximize the total reward of the allocation. In addition, the immediate reward is calculated using a dynamic programming (DP) algorithm based on predicted link states in advance because the link allocation problem can be treated as a 0-1 knapsack problem. The reward of the allocation to the beam is based on both the bandwidth and throughput per hertz of the user equipment (UE) and the allocation policies of the satellite operators.

The remainder of this chapter is organized as follows. In Section 5.2, the target SATCOM environment is described. In Section 5.3, the beam-series finite MDP for the frequency resource allocation problem is proposed to maximize the total reward using the Q-learning algorithm and the link allocation using the DP algorithm. In Section 5.4, the proposed resource allocation scheme is validated through the simulations. Finally, the chapter is concluded in Section 5.5.

5.2 Assumed Environment

In the following, we describe the SATCOM configuration of a bent-pipe HTS with the digital channelizer. Then, the system model is introduced.

5.2.1 System configuration

Figure 5.1 is an overview of the target SATCOM environment. In this system, a bent-pipe geostationary earth orbit (GEO) satellite provides multiple spot beam user links and multiple beam feeder links. In addition, the next-generation bent-pipe HTS is equipped with a digital channelizer, which enables flexible allocation of frequency resources and flexible beam assignments in response to fluctuating traffic requests. The digital channelizer has demultiplexers, switches, and multiplexers [4]. After being converted to digital signals using analog-to-digital converters, the input uplink signals are

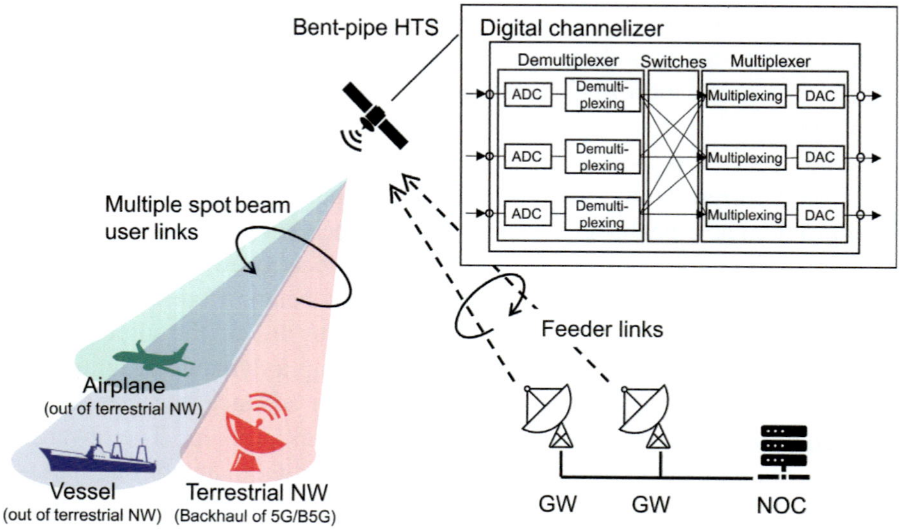

Figure 5.1 Overview of assumed SATCOM environment.

demultiplexed. Then, each demultiplexed signal is switched to the appropriate frequency, multiplexed, and processed through a digital-to-analog converter. The spectral efficiency is improved by the digital channelizer because each traffic request can be assigned to the desired frequency as long as no duplicated frequency resource is allocated in the adjacent beams.

The service types of nonterrestrial network (NTN) UE in the feeder links are classified into fixed satellite services ground stations, mobile satellite services including aircrafts, ships, and very small aperture terminals. NTN UE is equipped with modems that support the second-generation Digital Video Broadcasting by Satellite (DVB-S2) standard, in which multiple modulation and coding methods (MODCOD) can be selected. Note that UE at the edge of the beams suffers from lower carrier-to-noise ratios. In addition, the weather conditions around the UE are dominant because rain attenuation is strong in the Ka-band. Thus, the MODCODs differ depending on the given UE.

The satellite operation plans, including resource allocation in the transponder on the GEO-HTS, are managed in the network operations center (NOC). The satellite operator predicts the traffic demands and the channel states derived from the environmental conditions of the UE, for example, rain, clouds, and the distance from the supporting satellite. The prediction information is input to the NOC in advance and is utilized to plan the satellite operations. The HTS with the digital channelizer employs the Ka-band multibeam to enhance capacity and can support a variety of use cases, for example, the backhaul communication of a terrestrial network, nonterrestrial communication toward aircraft and ships, and disaster communication.

5.2.2 System model

Here, the frequency bandwidth in the downlink of the user links is denoted W MHz, the number of beams is denoted B, and the number of UE in beam b is denoted N_{b} as shown in Figure 5.2. The allocated frequency bandwidth in beam b is denoted a_{b} MHz and the sum of the allocated frequency bandwidth in the adjacent beams should not exceed the system bandwidth W MHz. Note that a frequency reuse method for satellite systems is described in TR 38.821 of Release 16 [9]. In the circularly polarized wave, a right-handed circularly polarized wave and a left-handed circularly polarized wave can be used selectively, and interference between the beams can be reduced considerably

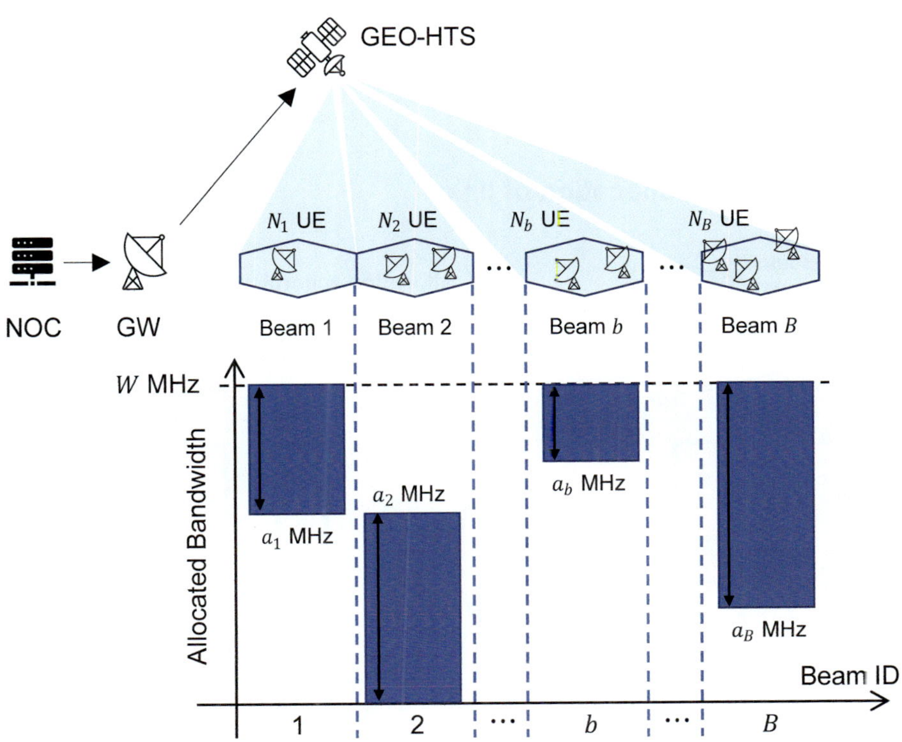

Figure 5.2 Frequency bandwidth allocated to HTS multibeam.

using different circularly polarized waves for the adjacent beam sequences. Thus, the bandwidth constraint $C1$ of the resource allocation to beam b is expressed as follows:

$$C1 : \begin{cases} a_{\mathrm{b}-1} + a_{\mathrm{b}} \leqq W \\ a_{\mathrm{b}} + a_{\mathrm{b}+1} \leqq W \end{cases} \text{, if } 2 \leqq \mathrm{b} \leqq B - 1. \tag{5.1}$$

Here, the request throughput of UE $\mathrm{u_b}$ in beam b is denoted $q^{\mathrm{TH}}_{\mathrm{u_b}}$ and defined as $q^{\mathrm{TH}}_{\mathrm{u_b}} = q^{\mathrm{BW}}_{\mathrm{u_b}} \eta_{\mathrm{u_b}}$, where $q^{\mathrm{BW}}_{\mathrm{u_b}}$ is the requested frequency bandwidth, and $\eta_{\mathrm{u_b}}$ is the throughput per hertz of UE $\mathrm{u_b}$.

5.3 Allocation Scheme of Frequency Resources

This section introduces the proposed frequency resource allocation scheme. First, we introduce the scheme used to allocate frequency resources to links using DP, and then we introduce the scheme used to allocate frequency resources to beams with the Q-learning algorithm. In addition, the immediate reward maximization scheme and a full search scheme are explained for reference.

5.3.1 Resource allocation of links

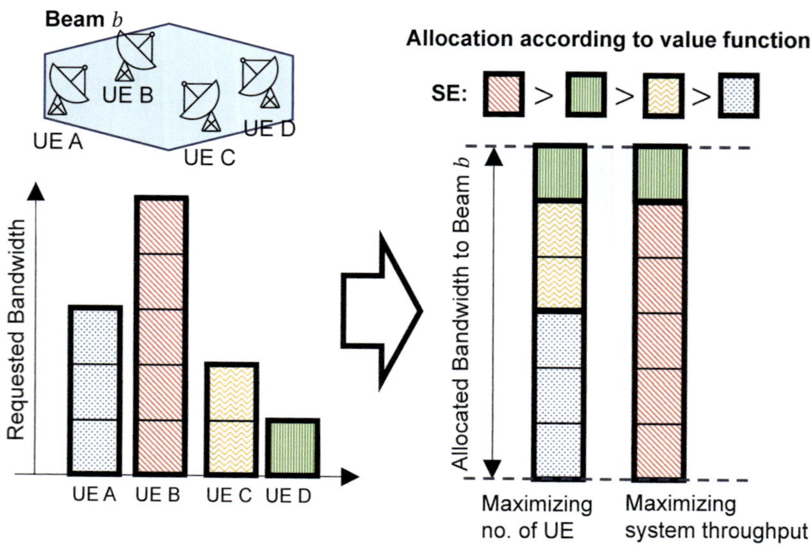

Figure 5.3 Example of link allocation.

As discussed in Section 5.2.1, different UE adopts different MODCODs according to the link states even if they belong to the same beam b. Note that the optimal allocation bandwidth to beam b is dependent on the evaluation criteria, as shown in Figure 5.3. The number of allocated UE is maximized when UE with a smaller requested frequency bandwidth is allocated to the links, which means that UE A, UE C, and UE D are allocated as shown in Figure 5.3. In addition, the system throughput is maximized when UE with larger SE values are allocated, which means that UE B and UE D are allocated as shown in Figure 5.3.

The problem of allocating frequency resources to UE under the condition that the allocated bandwidth ab can be treated as a 0-1 knapsack problem belongs to nondeterministic polynomial-time hardness.

$$\max \sum_{u_\mathrm{b}}^{\mathrm{N_b}} v_{u_\mathrm{b}} x_{u_\mathrm{b}}, \tag{5.2}$$

$$\text{subject to} \sum_{u_\mathrm{b}}^{\mathrm{N_b}} q_{u_\mathrm{b}}{}^{\mathrm{BW}} x_{u_\mathrm{b}} \leqq a_\mathrm{b},$$

$$x_{u_\mathrm{b}} \in \{0, 1\},$$

$$u_\mathrm{b} \in \{1, \cdots, \mathrm{N_b}\}.$$

Here, $x_{\mathrm{u_b}}$ represents whether UE $\mathrm{u_b}$ is allocated to the link and can take a value of 0 or 1. The weight $v_{\mathrm{u_b}}$ changes the characteristics of the system and is set as $v_{\mathrm{u_b}} = q_{\mathrm{u_b}}^{\mathrm{TH}}$ or $v_{\mathrm{u_b}} = 1$ to maximize the system throughput or maximize the number of allocated UE, respectively. To apply all combinations of UE allocated to the link, the complexity becomes $O(2^{N_\mathrm{b}})$. However, efficient solvers are well known in the 0-1 knapsack problem. In the target allocation problem, we adopt the DP shown in **Algorithm 5.1**, where the maximum total value up to UE $\mathrm{u_b}$ with the allocated bandwidth a_b is represented as $V[\mathrm{u_b}, \mathrm{a_b}]$. Here, the DP algorithm breaks down the link allocation problem into simpler subproblems, and the optimal solutions of the subproblems are determined recursively. The optimal combination $\boldsymbol{x}_{a_\mathrm{b}}^{opt}$ of the allocated UE under the allocated bandwidth a_b is determined according to **Algorithm 5.1**, where:

$$\boldsymbol{x}_{a_\mathrm{b}}^{opt} = \{x_1^{opt}, \cdots, x_{u_\mathrm{b}}^{opt}, \cdots, x_{N_\mathrm{b}}^{opt}\}. \tag{5.3}$$

Algorithm 5.1 Dynamic programming.

1: Acquire the information of the allocated bandwidth a_{b} to the beam b and the request bandwidth $q_{\mathrm{u_b}}^{\mathrm{BW}}$ and the allocation value $v_{\mathrm{u_b}}$ of each UE in the beam b

2: $u_{\mathrm{b}} = 1$

While $u_{\mathrm{b}} \leqq \mathrm{N_b}$

if $q_{\mathrm{u_b}}^{\mathrm{BW}} \geq a_{\mathrm{b}}$

3: $V[\mathrm{u_b}, a_{\mathrm{b}}] = V[\mathrm{u_b} - 1, a_{\mathrm{b}}]$

else

4: $V[\mathrm{u_b}, a_{\mathrm{b}}] = \max\left(V[\mathrm{u_b} - 1, a_{\mathrm{b}}], V[\mathrm{u_b} - 1, a_{\mathrm{b}} - q_{\mathrm{u_b}-1}^{\mathrm{BW}}] + q_{\mathrm{u_b}}^{\mathrm{BW}}\right)$

end if

5: $\mathrm{u_b} = \mathrm{u_b} + 1$

end while

5.3.2 Resource allocation of beams

Equation (5.1) indicates that the allocated frequency bandwidth in the previous beam determines the maximum allocation in beam b. Thus, resource allocation to the beams is considered to be a Markov process. In addition, this Markov process can be modeled as the beam-series finite MDP because the frequency interval of the digital channelizer is a discrete value. Here, the frequency bandwidth allocated in beam b, that is, a_{b}, is expressed as follows:

$$\begin{aligned} a_{\mathrm{b}} &= nI, \\ n &\in \left\{0, \cdots, \frac{W}{I}\right\}, \end{aligned} \tag{5.4}$$

where I denotes the frequency interval of the digital channelizer. The state S and the action A for beam b is defined as follows:

$$S : s_{\mathrm{b}} \in \left\{s^0, s^1, \cdots, s^n, \cdots, s^{\frac{W}{I}}\right\}, \tag{5.5}$$

$$A : a_{\mathrm{b}} \in \{0, I, \cdots, nI, \cdots, W\}, \tag{5.6}$$

where the state s_{b} in beam b is the frequency bandwidth allocated in previous beam $\mathrm{b} - 1$.

We employed the Q-learning algorithm to solve this finite MDP. If the action a_{b} is selected in beam b, and the state is transitioned from s_{b} to $s_{\mathrm{b+1}}$, $Q(s_{\mathrm{b}}, a_{\mathrm{b}})$ is updated as follows:

$$\begin{aligned} Q(s_{\mathrm{b}}, a_{\mathrm{b}}) \leftarrow &(1 - \alpha) Q(s_{\mathrm{b}}, a_{\mathrm{b}}) + \\ &\alpha \left(r_{\mathrm{b}} + \gamma \max_{a_{\mathrm{b+1}}} Q(s_{\mathrm{b+1}}, a_{\mathrm{b+1}}) \right), \end{aligned} \tag{5.7}$$

where α denotes the learning rate, γ denotes the discounting rate, and r_{b} denotes the immediate reward for action a_{b} in state s_{b}. Note that the immediate reward r_{b} depends on the policy of the NOC. The reward that maximizes the system throughput is given by the policy Φ_{TH}, and the reward that maximizes the number of allocated UE is given by the policy Φ_{UE}. For each condition, the reward r_{b} is expressed as follows:

$$\Phi_{\mathrm{TH}} : r_{\mathrm{b}} = \begin{cases} \sum_{u_{\mathrm{b}}}^{\mathrm{N_b}} q_{u_{\mathrm{b}}}^{\mathrm{TH}} x_{u_{\mathrm{b}}} - & \phi\left(a_{\mathrm{b}} - \sum_{u_{\mathrm{b}}}^{\mathrm{N_b}} q_{u_{\mathrm{b}}}^{\mathrm{BW}} x_{u_{\mathrm{b}}}\right) \\ & \textit{if } \boldsymbol{C}1 \textit{ is satisfied}, \\ -1 & \textit{if } \boldsymbol{C}1 \textit{ is } \textbf{\textit{not}} \textit{ satisfied}, \end{cases} \tag{5.8}$$

or

$$\Phi_{\mathrm{UE}} : r_{\mathrm{b}} = \begin{cases} \sum_{u_{\mathrm{b}}}^{\mathrm{N_b}} x_{u_{\mathrm{b}}} & \textit{if } C1 \textit{ is satisfied}, \\ -1 & \textit{if } C1 \textit{ is } \textbf{\textit{not}} \textit{ satisfied}, \end{cases} \tag{5.9}$$

where the immediate reward is -1 if the constraint given in eqn (5.1) is not satisfied.

Each beam has its own Q-table because the various conditions of each UE, for example, the number of UEs, the requested throughputs, and the MODCOD of each UE differ considerably in all beams. Figure 5.4 shows the update flows of the Q-table for beam b. First, the current action a_{b} is selected based on the current state s_{b} determined by the previous action $a_{\mathrm{b}-1}$ under the constraint that the same frequency resource cannot be assigned in adjacent beams. Second, the link allocation is performed using the DP algorithm in the allocated bandwidth a_{b} to beam b. Third, the reward r_{b} is calculated according to the policy. Finally, the Q-table is updated based on eqn (5.7). Assume that the number of episodes is denoted E, the number of reward calculations is given by EB.

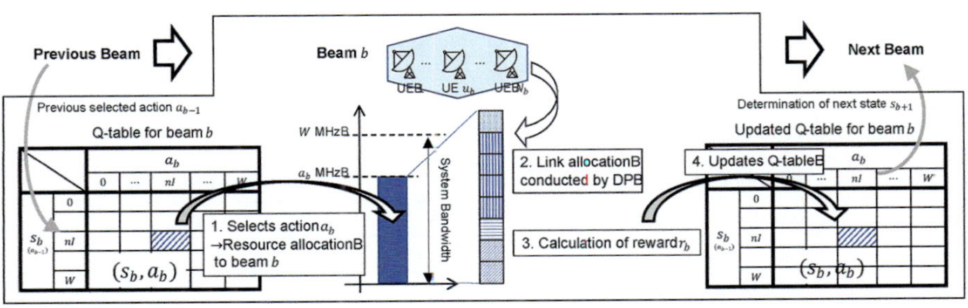

Figure 5.4 Update flow of Q-table for beam b.

Here, we define the immediate reward maximization scheme and the full search scheme for reference.

All possible actions in beam b are explored, and the action with the best immediate reward is selected in the immediate reward maximization scheme. The same selection is repeated for all beams. In this scheme, the number of reward calculations is given as $B\left(\frac{W}{I}+1\right)$.

In addition, all possible combinations of actions for all beams are explored, and the combination of actions with the best total reward is selected in the full search scheme. In this scheme, the number of reward calculations is expressed $\left(\frac{W}{I}+1\right)^B$.

5.4 Numerical Results

In this section, we discuss a computational simulation conducted to evaluate the proposed scheme.

5.4.1 Simulation conditions

The simulation parameters are described in Table 5.1. As can be seen, the system frequency bandwidth is set to 500 MHz, and the frequency interval of the digital channelizer is set to 50 MHz. In addition, the number of beams is set to 5, the maximum number of UE per beam is set to 5. The MODCOD parameter are shown in Table 5.2. Here, the number of actions for the frequency allocation to beams is determined as eqn (5.4) and (5.6), and the number of states is 11 according to eqn (5.5). The parameters of the Q-learning algorithm, that is, the learning and discounting rates, are set to 0.1 and 0.9, respectively, and the number of episodes is set to 3000.

Table 5.1 Simulation parameters.

Parameter	Value
System frequency bandwidth W	500 [MHz]
Frequency interval of digital channelizer I	50 [MHz]
Number of beams B	5
Maximum number of UEs per beam	5
MODCOD scheme	Table 5.2
Number of actions	11
Number of states	11
Learning rate α	0.1
Discounting rate γ	0.9
Number of episodes	3000

Table 5.2 Spectral efficiency according to MODCOD in DVB-S2 [10].

MODCOD	SE η [bps/Hz]	MODCOD	SE η [bps/Hz]	MODCOD	SE η [bps/Hz]
QPSK 2/9	0.434841	16APSK 3/5	2.370043	64APSK 32/45-L	4.206428
QPSK 13/45	0.567805	16APSK 3/5-L	2.370043	64APSK 11/15	4.338659
QPSK 9/20	0.889135	16APSK 28/45	2.458441	64APSK 7/9	4.603122
QPSK 11/20	1.088581	16APSK 23/36	2.524739	64APSK 4/5	4.735354
8APSK 5/9-L	1.647211	16APSK 2/3-L	2.635236	64APSK 5/6	4.936639
8APSK 26/45-L	1.713601	16APSK 25/36	2.745734	128APSK 3/4	5.163248
8PSK 23/36	1.896173	16APSK 13/18	2.856231	128APSK 7/9	5.355556
8PSK 25/36	2.062148	16APSK 7/9	3.077225	128APSK 29/45-L	5.065690
8PSK 13/18	2.145136	16APSK 77/90	3.386618	128APSK 2/3-L	5.241514
16APSK 1/2-L	1.972253	32APSK 2/3-L	3.291954	256APSK 31/45-L	5.417338
16APSK 8/15-L	2.104850	32APSK 32/45	3.510192	256APSK 32/45	5.593162
16APSK 5/9-L	2.193247	32APSK 11/15	3.620536	256APSK 11/15-L	5.768987
16APSK 26/45	2.281645	32APSK 7/9	3.841226	256APSK 3/4	5.900855

5.4.2 Performance evaluation

Figure 5.5 shows the average system throughput relative to the number of episodes for the Q-learning algorithm when the NOC adopts a policy that maximizes the system throughput. Here, the system throughput with the full search scheme is represented by the blue line, and that with the immediate reward maximization scheme is represented by the green line. In addition, the system throughput with the Q-learning algorithm is represented by the orange line. As can be seen, this line approaches the maximum value as the number of episodes increases. The probability of reaching the maximum throughput versus the number of episodes for the Q-learning algorithm is shown in Figure 5.6. This represents the probability that the solution of the Q-learning algorithm and the immediate reward maximization scheme reaches that of the full search scheme. The probability of reaching the maximum throughput with the Q-learning algorithm is approximately 50% when the number of episodes is 3000, and with the immediate reward, the maximization scheme is less than 20% because this scheme only reaches the optimal solution when the sum of the requested frequency in all adjacent beams is less than the system bandwidth, that is, 500 MHz. In contrast, the Q-learning algorithm is more inclined to find the optimal solution because the solution space is efficiently explored since the action in which the Q value updated based on the learning rate is relatively high is centrally explored when the balance between exploration and exploitation is appropriate. On the other hand, the Q-learning algorithm is more inclined to find the optimal solution because it efficiently explores the solution space when the balance of exploration and exploitation is appropriate.

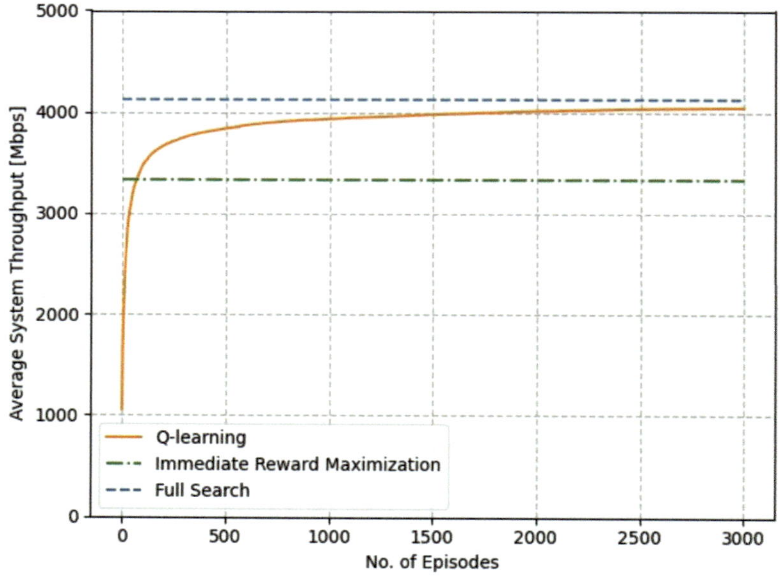

Figure 5.5 Average system throughput versus number of episodes.

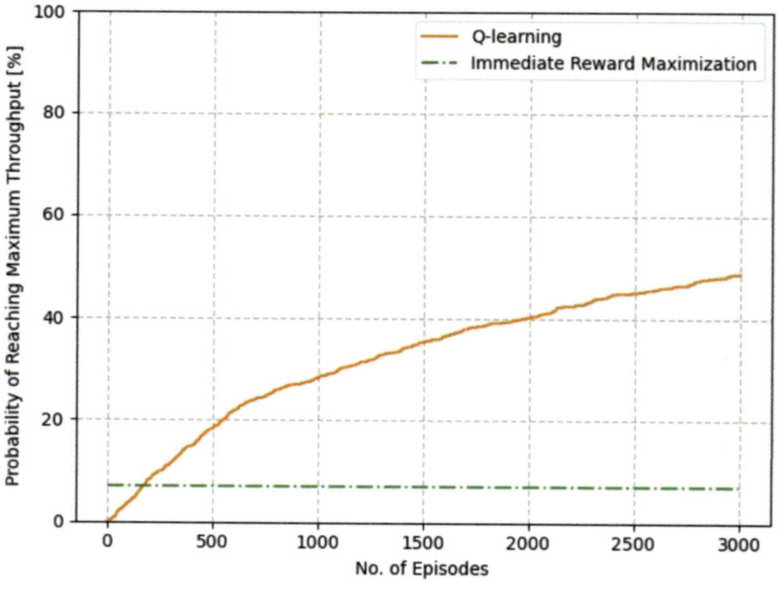

Figure 5.6 Probability of reaching maximum throughput vs. number of episodes.

Figure 5.7 shows the distribution of the system throughput of the immediate reward maximization scheme and the Q-learning algorithm for the optimum solution obtained by the full search. Note that the full search explores the entire solution space and reaches the optimal beam allocation while the other schemes do not always realize the optimal allocation. As can be seen, the optimal solution obtained by the full search is arranged on a straight line as an upper bound, and the system throughput values with the other schemes are below the line. The average system throughput obtained with the full search is 4127 Mbps, that obtained with the Q-learning algorithm is 3902 Mbps, and that obtained with the immediate reward maximization scheme is 3335 Mbps. The average deterioration rate of the Q-learning algorithm compared to the full search is 5.39%, and that of the immediate reward maximization is 20.5%. Here, the deterioration rate indicates the percentage of degradation from the optimal solution obtained by the full search. The standard deviation for the deterioration rate of the Q-learning algorithm is 5.37%, and that of the immediate reward maximization is 16.7%. These results indicate that frequency resource allocation with the Q-learning algorithm has a solution distribution that is closer to the optimum solution than that with the immediate reward maximization scheme.

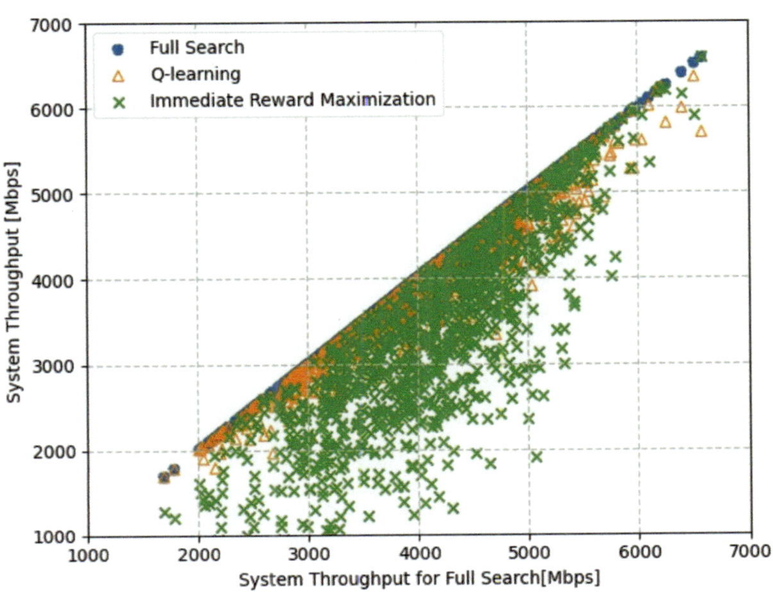

Figure 5.7 Distribution of system throughput.

Figures 5.8 shows the average number of allocated UE versus the number of episodes and Figure 5.9 shows the probability of reaching the maximum number of allocated UE versus the number of episodes when the NOC adopts the policy that maximizes the number of allocated UE. The average number of allocated UE with the Q-learning algorithm is closer to that of the full search scheme when the number of episodes is 3000. In addition, as shown in Figure 5.8, convergence to the maximum number of allocated UE is faster than that of the system throughput shown in Figure 5.5. In support of this observation, the probability of reaching the maximum number of allocated UE with the Q-learning algorithm is greater than 80% when the number of episodes is 3000. This is because the large number of MODCODs shown in Table 5.2 that can be adopted by each UE results in a complicated allocation problem when the NOC policy maximizes the system throughput while the number of allocated UE calculated from eqn (5.9) is the sum of the variables x_{u_b} represented by 0 or 1 in eqn (5.3) when the NOC policy maximizes the number of allocated UE.

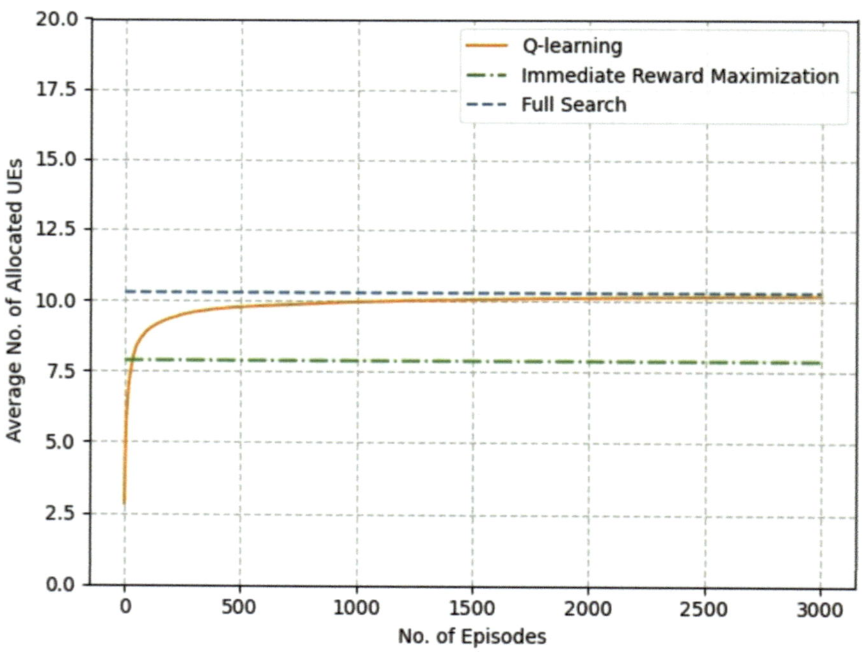

Figure 5.8 Average number of allocated UEs versus number of episodes.

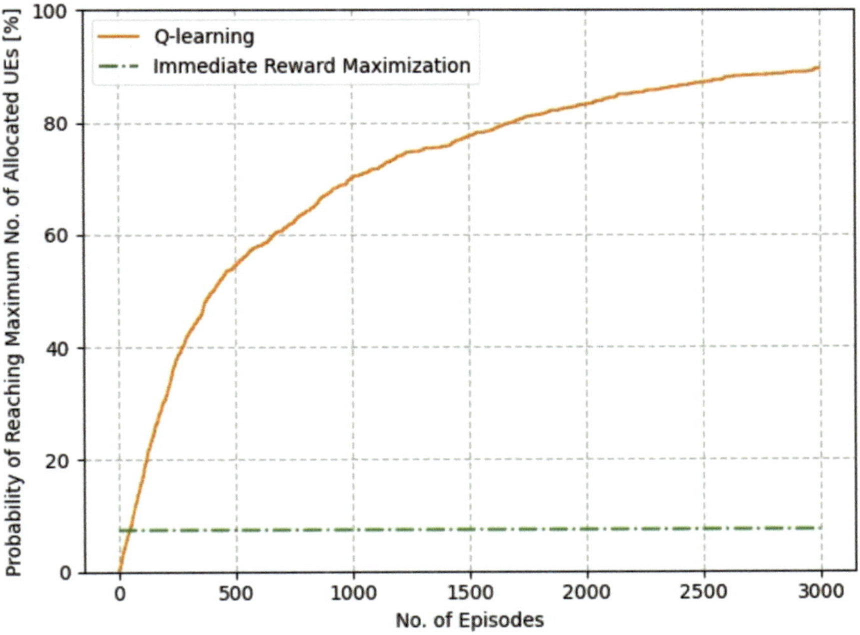

Figure 5.9 Probability of reaching maximum number of allocated UEs versus number of episodes.

Figure 5.10 shows the distribution of the system throughput of the immediate reward maximization scheme and the Q-learning algorithm for the optimum solution obtained by the full search strategy. Here, compared to Figure 5.7, the solution distribution is discrete because the NOC adopts the policy Φ_{UE} represented by eqn 5.9, and the immediate reward r_{b} is given by the integer values. The average number of allocated UE with the full search is 10.3, that with the Q-learning algorithm is 9.9, and that with the immediate reward maximization scheme is 7.9, as shown in Figure 5.8. The average deterioration rate of the Q-learning algorithm compared to the full search strategy is 3.25%, and that of the immediate reward maximization is 23.3%. The deterioration rate indicates the percentage of degradation from the optimal solution obtained by the full search strategy. The standard deviation for the deterioration rate of the Q-learning algorithm is 4.86%, and that of the immediate reward maximization is 12.9%. The standard deviation of the number of allocated UEs is lower than that of the system throughput.

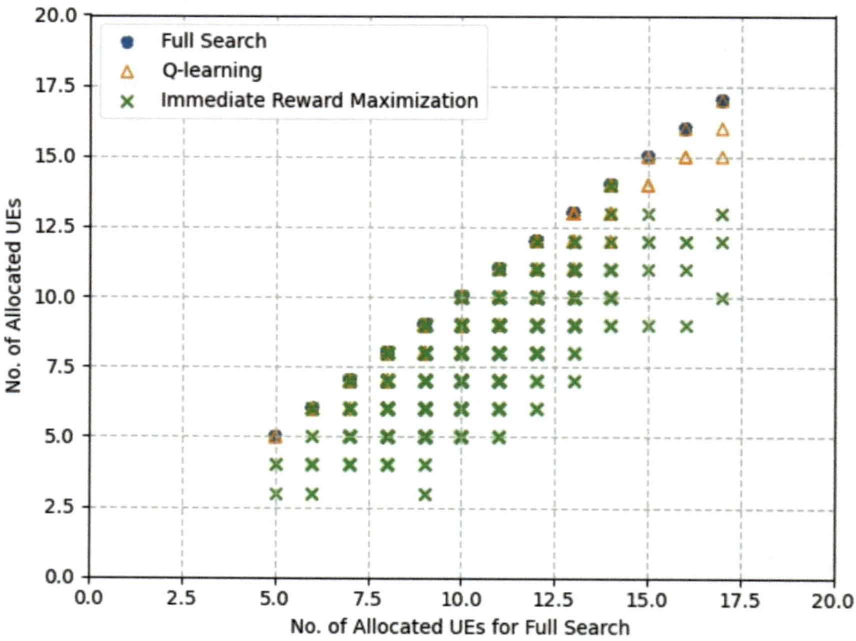

Figure 5.10 Distribution of number of allocated UEs.

The number of link allocation combinations per beam is the same for Φ_{UE} and Φ_{TH} in 2^N ways. However, in Φ_{UE}, the value that can be taken by the allocation of a single UE is binary, as indicated by eqn (5.9), whereas in Φ_{TH}, the throughput and loss for the allocated bandwidth are multivalued, as indicated by eqn (5.8). Thus, the reward values in Φ_{TH} are greater than those in Φ_{UE}, and the solution space is more expanded. This is why the convergence characteristic of Φ_{TH} is slower than that of Φ_{UE}, and the standard deviation of the degradation rate for the entire search is large.

5.5 Conclusions and Future Works

This study proposed a scheme to allocate frequency resources to beams using the Q-learning algorithm in HTS communication systems. Here, the resource allocation problem was modeled as a beam-series finite MDP, and the frequency resource allocation to links was solved using a DP algorithm. In addition, the Q-table was updated based on the reward. The performance and the complexity of the proposed resource allocation scheme were compared with the full search and immediate reward maximization schemes.

Our future task is adopting a deep Q-learning or a state-of-art AI algorithm to approximate Q-value from untrained data when the frequency interval of the digital channelizer becomes more fine-grained and the entire solution space is expanded.

References

[1] Satellite Industry Association, "State of the Satellite Industry Report 2022." Available: https://sia.org/news-resources/state-of-the-satell ite-industry-report/

[2] J. Cherkaoui and V. Glavac, "Signal Frequency Channelizer/ Synthesizer," 2008 10th International Workshop on Signal Processing for Space Communications, 2008.

[3] H. Yang, J. Dang, Y. Pan and Z. Li, "A Digital Channelizer Design Approach for Broadband Satellite Communications Based on Frequency Domain Filter Theory," Proceedings 2013 International Conference on Mechatronic Sciences, Electric Engineering and Computer (MEC), pp. 2986-2990, 2013.

[4] K. Kaneko, H. Nishiyama, N. Kato, A. Miura and M. Toyoshima, "Construction of a Flexibility Analysis Model for Flexible High-Throughput Satellite Communication Systems With a Digital Channelizer," in IEEE Transactions on Vehicular Technology, vol. 67, no. 3, pp. 2097-2107, March 2018.

[5] Y. Abe, H. Tsuji, A. Miura and S. Adachi, "Frequency Resource Allocation for Satellite Communications System Based on Model Predictive Control and Its Application to Frequency Bandwidth Allocation for Aircraft," 2018 IEEE Conference on Control Technology and Applications (CCTA), pp. 165-170, 2018.

[6] Y. He, B. Sheng, H. Yin, D. Yan and Y. Zhang, "Multi-objective deep reinforcement learning based time-frequency resource allocation for multi-beam satellite communications," in China Communications, vol. 19, no. 1, pp. 77-91, Jan. 2022.

[7] D. Zhao, H. Qin, N. Xin and B. Song, "Flexible Resource Management in High-Throughput Satellite Communication Systems: A Two-Stage Machine Learning Framework," in IEEE Transactions on Communications, vol. 71, no. 5, pp. 2724-2739, May 2023.

[8] G.Otsuru, Y. Sanada, H. Tsuji, and A. Miura, "Frequency Resource Allocation Using Q-learning for User Link in High Throughput Satellite Communication," The 26th International Symposium on Wireless Personal Multimedia Communications, Nov. 2023.

[9] "Technical Specification Group Radio Access Network; Solutions for NR to support non-terrestrial networks (NTN)," 3GPP, TR 38.821, V16.20.0, Mar. 2023.

[10] ETSI, "Digital Video Broadcasting (DVB); Implementation Guidelines for the Second Generation System for Broadcasting, Interactive Services, News Gathering and Other Broadband Satellite Applications; Part 2: S2 Extensions (DVB-S2X)", ETSI TR 102 376-2 V1.2.1.

6

K-medoids and BIRCH Clustering Methods for Analyzing 6G High MmWave Bands

Saud Alhajaj Aldossari

Department of Electrical Engineering, College of Engineering in Wadi Alddawasir, Prince Sattam Bin Abdulaziz University, Saudi Arabia
E-mail: S.alhajaj@psau.edu.sa

Abstract

Since the early stages of advanced mobile phone systems (AMPS), in the first generation (1G), mobile systems have significantly evolved. Today, with 5G and beyond, big data and artificial intelligence (AI) have created a new era in communication systems. This has led to the research and development of 6G wireless communication, which will become available in the early 2030s. 6G wireless communications will overcome the drawbacks of previous generations using high mmWave and terahertz (THz) frequencies combined with AI. This chapter presents an analysis of clustering models for binary variable analysis. The objective is to evaluate the performance of these models and gain insights into the underlying data patterns of the 145 GHz frequency in an urban microenvironment. The selected 145 GHz band falls within the millimeter wave (mmWave) band, which offers an extremely high bandwidth compared to those at lower frequencies. This enables the transmission of large amounts of data, making it effective for applications such as high-speed data links and future generations of wireless communication (e.g., 6G). Moreover, this frequency has not been investigated because the signal attenuation (weakening) is greater at these frequencies, limiting the transmission range. K-medoids and BIRCH clustering techniques are used to understand the relationship between channel state information (CSI). Using metrics and visualizations, the author aims to provide a comprehensive understanding

DOI: 10.1201/9788743801108-7

of the data and facilitate informed decision-making. K-medoids and BIRCH clustering techniques have been used to understand the relationship between CSI.

Keywords: 5G, 6G, artificial intelligence, clustering, K-medoids, BIRCH

6.1 Introduction

Artificial intelligence (AI) techniques, such as clustering, are used to enhance the understanding of wireless communications based on the provided data [1, 2]. This study investigates higher mmWave bands using clustering methods such as BIRCH and K-medoids to show the relationship between the channel state information (CSI) parameters and the development of clustering methods that are both accurate and efficient. Developing new clustering methods for 6G wireless communications is an active area of research. Study [3] explored channel multipath components using clustering. In contrast, [4] presented the challenges and opportunities of using clustering methods in wireless channel modeling. Furthermore, [5] tracked time and variant cluster parameters in MIMO systems. Study [6] performed clustering algorithms for UAV networks. 6G wireless communication is a promising technology expected to provide significantly higher data rates and lower latency than previous generations of wireless networks [7]. The new behavior of the channel state parameters leads to the learning and investigation of these behaviors. Currently, it is necessary to understand the propagation channel of the new era of big data and new systems, such as 6G, and how it works with higher-frequency bands.

Clustering is an unsupervised learning technique that can be used to group unlabeled data points based on their similarities [8]. This can be useful for identifying patterns and trends in wireless data for several reasons. This information can then be used to make better decisions on how to operate a network. In the clustering phase, we explored two models: K-medoids and BIRCH [9, 10]. Subsequently, evaluation metrics such as the silhouette score, Calinski–Harabasz score, and Davies–Bouldin score were utilized to measure the quality of the clusters [11, 12]. Visualizations, including cluster proportions and scatter plots, compare features to enhance the interpretation of the results. Using metrics and visualizations enables a thorough evaluation of the model performance and enhances our understanding of binary variable analysis.

The primary contributions of this study are as follows:

- It presents a brief overview of the state-of-the-art 6G.
- It demonstrates how AI enhances wireless channel modeling by improving accuracy and efficiency in predicting channel characteristics.
- It proposes clustering methods to support the 6G band investigation.
- It provides a comprehensive understanding of the data and facilitates informed decision-making.

The rest of this chapter is organized as follows. Section 6.3 explores the channel modeling of 6G. Section 6.4 addresses the knowledge of AI. The results, discussion, and future work are included in Sections 6.5 and 6.6. Finally, the conclusions are presented in Section 6.7.

6.2 6G Radio

6G radio, the next generation of wireless technology, is still in the early stages of development and has been empowered by numerous different innovations [13]. The presence of 5G New Radio (NR) networks in current and cutting-edge remote network cases is an essential challenge that should be considered. NR works in the mmWave band, and beamforming is a significant innovation for upgrading gains. Beamforming is fundamental for expanding the link budget in the bandwidth of mmWave correspondence. 6G is superior to existing networks in empowered administrations and applications. These networks support excessive client demands for high speed, reduced latency, and greater consistency. To meet these demands, the 6G architecture provides versatility and can enable diverse applications. Compared with previous eras, distinct new elements in 6G have been created to be adaptable to ensure that new functions may be incorporated. The adaptability of 6G originates from its adaptable spectrum and flexible numerology. Different 6G received signals contain various recurrence bands with the RIS and other signals, such as those from indoor IoT devices. 6G, mostly, uses higher frequency bands. It uses a large transmission bandwidth to enable a high-speed information rate and traffic limitations. Furthermore, 6G uses antenna arrays to transmit and receive data. Thereby, 6G networks support many devices and provide high-speed data rates to each device. Moreover, 6G uses AI and machine learning (ML) to improve performance and efficiency. AI and ML can be used to optimize spectrum allocation, manage traffic, and improve the reliability of 6G networks [14].

The 6G evolution must address creative challenges and overcome various nonspecialized aspects, most notably industry hindrances, range distribution, and the use of AI with specified guidelines. Unlike 5G, 6G will permeate all friendly creations and living viewpoints and be more firmly organized into different undertakings. Versatile will not be restricted to a specific field but will collaborate intimately with different ventures/fields. 6G THz networks will require coordinating tasks across different regions and countries globally to ensure seamless integration and performance. The frequency range should be as broad as possible to ensure optimal signal coverage. Coordination with other researchers regarding climate radars should also be considered [15]. Satellite communication faces increasing constraints in terms of plans and rules. In addition, satellite correspondences have more issues trading overall meandering than traditional natural interchanges. Currently, a few nations and businesses are building satellite communication frameworks using AI. Coordinating the connections between these independently deployed satellite communication systems will be a complex challenge.

6.3 Channel Modeling in 6G

Channel modeling is a method used to create a mathematical representation of a wireless communication channel. A wireless channel is inherently variable depending on the environment, which is the rationale for several distinct models being developed. Typically, these models are used to simulate the effect of the channel on the transmitted signal and to predict the performance of 6G systems. THz waves have higher frequencies than micro and millimeter waves. Researchers are exploring the use of extremely high-frequency bands (above 100 GHz) for 6G communications because these frequencies provide a wider bandwidth and faster data transmission rates. However, THz signals have a considerably short range and can be easily blocked, making the deployment of 6G networks challenging. Furthermore, quantum communication using AI can offer high levels of security and privacy in 6G networks [16]. This is achieved using the principles of quantum mechanics to ensure that communication cannot be intercepted or tampered with by a third party. Moreover, 6G is immune to hacking and eavesdropping, making it a potential research area for 6G standards.

The propagation of 6G is expected to differ from that of previous wireless generation channels. This is because 6G uses higher frequencies, and its channels are prone to attenuation and blockage. Moreover, the 6G wavelength is shorter, governed by the following equation:

$$\lambda = \frac{c}{f}, \tag{6.1}$$

where λ is the wavelength, c is the speed of light, and f is the frequency. Therefore, the channel has higher temporal variation and contains more multipath components. Thus, to accurately model a 6G channel, the channel model should consider the following factors:

- Frequency: the frequency of a signal affects the attenuation and dispersion of the signal.
- Environment: the environment in which signals are sent and received also affects signal attenuation and dispersion.
- Mobility: transmitter and receiver mobility affects the temporal variation of the channel.

Several different channel modeling techniques are available for 6G. Some of the most common techniques include the following:

- Empirical channel models: models that are based on real measurements of channels in different environments [17].
- Stochastic channel models: these models use statistical distributions to represent various characteristics of the channel, and they include correlation-based stochastic models (CBSM) and geometry-based stochastic models (GBSM) [18].
- Deterministic channel models: these models use mathematical expressions to describe signal propagation through channels [19].

Hence, channel modeling is an important part of 6G network development. By accurately modeling the channel, we can better understand the performance of 6G systems, thereby making them more robust and efficient.

6.4 Artificial Intelligence in Wireless Communications

Recent advancements in AI research have significantly accelerated its application across various academic and commercial sectors. AI can be quickly used for cognitive radios, distance detection, and executive organization in communication and transmission handling [20]. Furthermore, in wireless communications, AI and its linked computations are constantly displaying value in various development processes, such as massive MIMO communications, that require productive channel evaluation and image recognition.

These assignments ordinarily do not offer low-intricacy ideal plans in convoluted channels to ensure equal handling inherent in AI can be effectively used to enhance computational limits. Current remote networks have a layered design in which each layer principally serves a couple of capacities, including computer-based intelligence. Large-scale computations dynamically bridge gaps between layers, enhancing overall performance across the network's wireless connections. This aspect is synchronized by considering the current layers, which offers a clear approach to exploring numerous inevitable computer-based intelligence applications. Simulated intelligence incorporates a few branches of science, such as mathematical models, normal language handling, AI, virtual reality (VR), augmented reality (AR), extended reality (ER), and PC vision. AI and ML are expected to manage 6G networks more efficiently and provide new services. AI and ML can be used to optimize network performance, predict network problems, and personalize user experiences.

6.4.1 Artificial intelligence in the physical layer

The physical layer has been modeling a preparation in which mathematical models with a particular design are proposed and updated inside the imperatives to meet the movement of predetermined execution necessities. For instance, a channel model can be used in relation to other parametric for direct channel assessment. These model-based frequently perform well if the enlistment of mathematical models is moderately evident [21, 22].

6.4.2 Artificial intelligence in wireless networks

The availability of rich data sets at other critical levels of wireless organizations suits the importance of AI-based arrangements. Specialists have successfully implemented support learning strategies in directing conventional plans for sensor organizations to obtain a more energy-efficient control arrangement for lower-sensor organizations [23]. For vehicular communication systems, predictive models based on continuous data are more accurate than traditional theoretical models due to constant advancements in vehicles. In addition to advancements in computational enhancement, a possible difficulty focuses on information storage and management, typically on cloud-based central server farms [24]. This substantially delays the accomplishment of tasks that require constant action by devices scattered across an extensive geographic range. Furthermore, a unified figuring method is not required for information security or handling restrictions. To address these issues, edge AI directs neighborhood devices to complete activities and administrative tasks [25]. This could create a conceivable overburden in neighborhood devices, as they are not equipped with as solid taking care of units as the cloud taking care of concentration.

6.4.3 Artificial intelligence in network administration

Intelligence network administration is essential for managing organizational operations and improving efficiency. With the widespread adoption of Software-Defined Networking (SDN) and Network Function Virtualization (NFV), the process of acquiring large-scale data has become significantly more streamlined and manageable [26]. This has presented significant areas of strength for AI-based administration and coordination natives inside 6G, ultimately prompting complete computerization, as discussed in the following section [27].

During the early stages of 5G standardization, experts investigated the potential of AI in accomplishing high time and reaching viability once solidified in 5G. Specifically, AI estimations can assist with the accompanying undertakings that would provide low-efficiency regular courses of action: recognizing network irregularities, appointing network resources, and performing network management, among others. However, these arrangements have not been fully embraced in 5G proposals worldwide. Previously, 5G networks relied on sub-6 GHz bands, and mmWaves were introduced owing to the need for high data throughput. The 6G group is intended to focus on THz frequencies to enable information speeds of up to Tbit/s [15]. Many enterprises transmit data via optical wireless communication (OWC). Although artificial consciousness in wireless communication will undoubtedly bring about a shift in perspective toward information-based techniques, unresolved challenges remain. No consensus has been reached on which computations work best to resolve a united issue in small, such as coding design, channel assessment, and resource segmentation. Practically distributed methods promise high precision or low complexity, with either logical hypotheses or valuable information collections.

The methodology of this work was conducted based on an unsupervised learning method called clustering. Channel estimation in 6G candidate frequency bands is complex, involving a combination of signal processing techniques and machine learning algorithms. Data generation was initiated via an open-source MATLAB simulation developed by New York University [28, 29] for high MmWave. Various CSI were generated, such as T-R separation distance (m), time delay (ns), received power (dBm), phase (rad), azimuth AoD (degree), elevation AoD (degree), azimuth AoA (degree), elevation AoA (degree), path loss (dB), and RMS delay spread (ns). Subsequently, data cleansing methods were conducted on the data set, such as removing duplicate or irrelevant observations to enhance the accuracy through validation and enrichment. Subsequently, unsupervised learning methods were conducted via Python to learn the behavior of high MmWave.

6.5 Results and Discussion

This section aims to analyze 6G CSI using the BIRCH and K-medoids clustering methods to demonstrate how AI can be used to assist the new generation of 6G. Using clustering, we can visualize the data, and based on this, decisions can be made. The data were generated for an urban microenvironment at a frequency of 145 GHz. The main feature of the CSI used was path loss; other details can be obtained from [30]. Furthermore, two clusters were created based on the path loss of the propagated signal. Thus, $PL \geq 120 \implies$ class zero. $PL < 120 \implies$ class one as shown in Figure 6.1. The maximum path loss value that allows the receiver to detect and decode the transmitted signal is assumed to be 120 dB; otherwise, the signal is considered to be noise [31]. The maximum allowable path loss typically depends on multiple factors, such as the transmitter, transmitter antenna gain, receiver sensitivity, receiver antenna, distance between the transmitter and receiver, frequency, and other system constants.

6.5.1 K-medoids

The K-medoids algorithm is a clustering method that partitions data into K clusters, where each cluster is represented by a medoid, which is the center

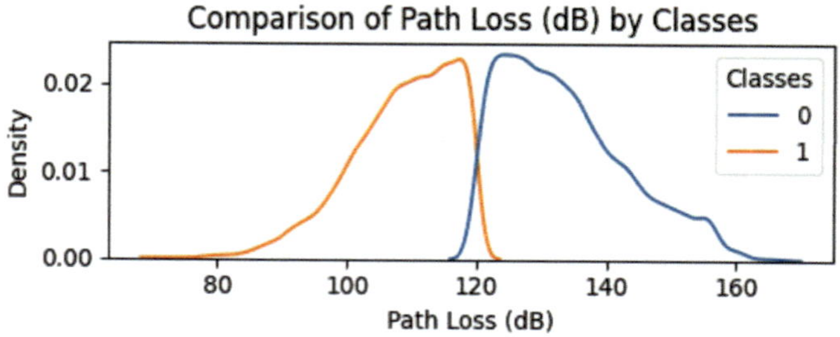

Figure 6.1 Classes Based on Path Loss.

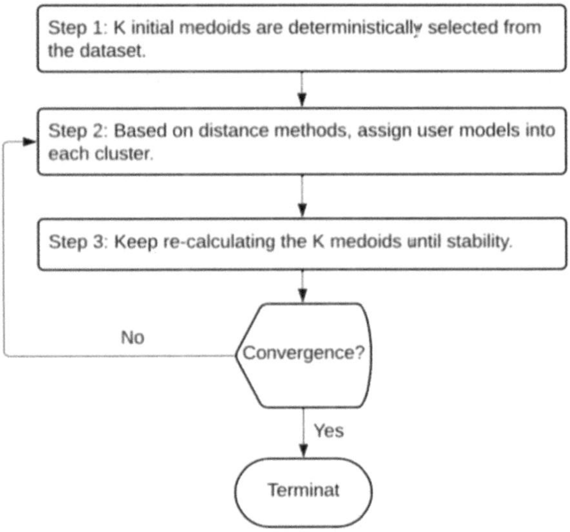

Figure 6.2 Flowchart of Unsupervised K-medoids.

Algorithm 6.1 K-medoids algorithm.

1: Initialization: Let $X = x_1, x_2, , ..., x_K$ be the set of medoids.
2: Assignment of points to medoids: For each data point x_i, the algorithm assigns it to the nearest medoid based on the dissimilarity or distance measure.
3: Distance measures, such as Manhattan distance, Chebyshev distance, and cosine similarity, can be calculated as follows: $c_i = argmin_j \sum_{x_k \in C_j} d(x_i.x_k)$
4: Update medoids, selecting the data point that minimizes the total dissimilarity within the cluster as a new medoid.
5: The previous steps are repeated until a convergence criterion is met, such as the stability of the medoids or the maximum number of iterations.

point of a cluster [32]. Unlike K-means, which uses centroids, K-medoids select actual data points as representatives of the clusters. The K-medoids algorithm is presented in the flowchart Figure 6.2 and Algorithm 6.1.

Algorithm 6.1 shows the procedure for K-medoids to cluster data, where all points are assigned to medoids. This algorithm attempts to improve clustering by considering alternative medoids within each cluster. It selects a data point that minimizes the total dissimilarity within the cluster as a new medoid. In step 2, C_j represents the cluster with medoid x_j, and $d(x_i, x_k)$ is the dissimilarity or distance between x_i and x_k. The K-medoids algorithm seeks to determine the optimal configuration of medoids that minimizes the total dissimilarity within each cluster. It is particularly useful when dealing with categorical or non-Euclidean data because it does not rely on calculating means. However, this can be computationally expensive, particularly for large data sets, because it requires computing the dissimilarities between all pairs of data points. The results of the K-medoids algorithm are listed in Table 6.1.

A clustering plot of K-medoids can be observed in Figure 6.3, which represents the cluster of both classes, and negative and positive scales are presented to show the overlap between the classes. As the distance increases, the path loss increases, and the received signal gradually decreases, which matches the Friis Transmission Formula that mainly states [33].

$$P_{\mathrm{R}} = \frac{P_{\mathrm{T}} G_{\mathrm{T}} G_{\mathrm{R}} C^2}{(4\pi RF)^2}, \tag{6.2}$$

where P_{R} is the received power and P_{T} is the transmitted power. While G_{R} and G_{T} are the transmitted and received antenna gains. R is the distance between the two antennas and C and F are the speed of light and frequency, respectively.

To assess the performance of the clustering algorithm, the following metrics were computed.

Table 6.1 K-medoids model results.

	Silhouette score	Calinski–Harabasz score	Davies–Bouldin score
K-medoids	0.15	5445.21	2.24

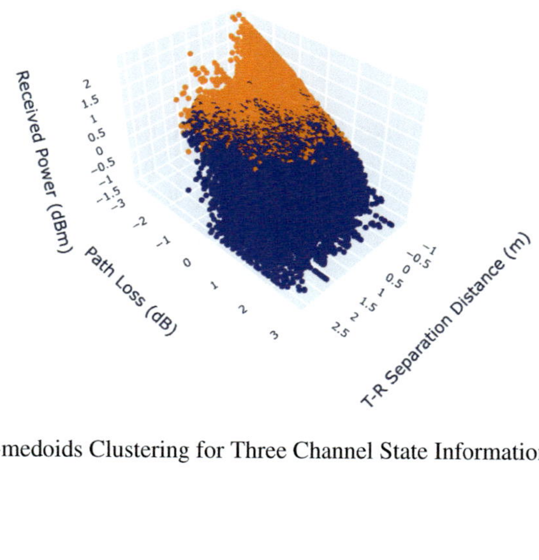

Figure 6.3 K-medoids Clustering for Three Channel State Information (CSI) Features.

The silhouette score measures the degree of separation between clusters. A higher score indicates better-defined clusters.

The Calinski–Harabasz score evaluates the ratio of between-cluster to within-cluster dispersions. Higher scores suggest well-separated and compact clusters.

The Davies–Bouldin score measures both the separation and compactness of clusters. A lower score indicates better clustering results.

6.5.2 BIRCH

Balanced Iterative Reducing and Clustering using Hierarchies (BIRCH) is an efficient hierarchical clustering algorithm designed for large-scale data sets [34]. It utilizes a tree-based structure called a clustering feature tree (CFT) to organize the data. This algorithm involves two main steps: clustering feature construction (CF Construction) and clustering feature selection (CF Selection). The CF Construction step involves the following equations:

- Micro-cluster centroid update:

$$\mathrm{CF_i}.N = \mathrm{CF_i}.N + 1\mathrm{CF_i}.\mathrm{LS} = \mathrm{CF_i}.\mathrm{LS} + x\mathrm{CF_i}.\mathrm{SS} = \mathrm{CF_i}.\mathrm{SS} + x^2, \tag{6.3}$$

  where $\mathrm{CF_i}$ represents the micro-cluster i, N is the number of data in the micro-cluster, LS is the linear sum of the points, x is a data point, and SS is the squared sum of the points. This equation is also part of the BIRCH micro-cluster centroid update process. The centroid is a representative point of the cluster, often calculated as the mean of the data points where it deals with updating three key aspects of the micro-cluster:

  - CF$i.N$: this represents the number of data points currently assigned to the micro-cluster i.
  - CFi.LS: this represents the linear sum of the data points in micro-cluster i. Essentially, it is the sum of all the data points' values for each dimension.
  - CFi.SS: this represents the sum of squares of the data points in micro-cluster i. It is the sum of the squares of each data point's value for each dimension.

- Micro-cluster radius update:

$$\mathrm{CF_i}.R = \sqrt{\frac{\mathrm{CF_i}.\mathrm{SS}}{\mathrm{CF_i}.N} - \left(\frac{\mathrm{CF_i}.\mathrm{LS}}{\mathrm{CF_i}.N}\right)^2}. \tag{6.4}$$

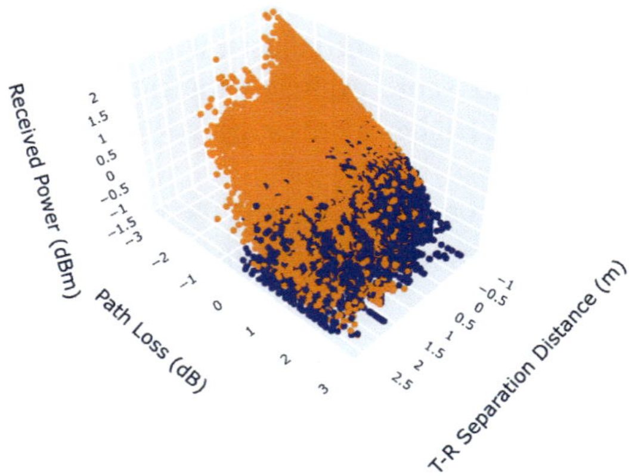

Figure 6.4 BIRCH Clustering for Three CSI Features.

The CF Selection step involves the following equation for merging two microclusters ($\mathrm{CF_i}$ and $\mathrm{CF_j}$):

$$\mathrm{CF_i}.N = \mathrm{CF_i}.N + \mathrm{CF_j}.N, \tag{6.5}$$

$$\mathrm{CF_i.LS} = \mathrm{CF_i.LS} + \mathrm{CF_j.LS}, \tag{6.6}$$

$$\mathrm{CF_i.SS} = \mathrm{CF_i.SS} + \mathrm{CF} + j.\mathrm{SS}. \tag{6.7}$$

Finally, a CFT is constructed based on these microclusters to enable efficient hierarchical clustering. BIRCH micro-cluster centroid update equations provide a mechanism for efficiently maintaining and updating micro-clusters in the algorithm, enabling efficient and accurate clustering of large data sets.

The BIRCH algorithm offers advantages such as scalability for large data sets, handling of high-dimensional data, and the ability to handle online data insertion. It is a valuable tool for various clustering tasks in which efficiency and scalability are paramount. BIRCH was used to study the CSI of a wireless channel, and three features were used, as shown in Figure 6.4

Based on the results obtained by comparing the cluster model metrics, the evaluation can be summarized as follows:

K-medoids: the K-medoids model performs reasonably well in categorizing binary variable classes. With a silhouette score of 0.15, it demonstrates

Table 6.2 BIRCH model results.

	Silhouette score	Calinski–Harabasz score	Davies–Bouldin score
BIRCH	0.14	3890.32	2.39

a moderate level of cluster cohesion and separation. The Calinski–Harabasz score of 5445.21 suggests good intercluster separation and intracluster similarity, supporting its effectiveness in distinguishing the two binary categories as shown in Figure 6.3. Additionally, the Davies–Bouldin score of 2.24 indicates a moderate level of cluster quality for "Classes."

BIRCH: the BIRCH model shows moderate performance in categorizing the binary variable "classes". From Table 6.2, a silhouette score of 0.14, indicates a reasonable level of cluster cohesion and separation. The Calinski–Harabasz score of 3890.32 suggests low intercluster separation and intracluster similarity, supporting its potential to distinguish the two binary categories, as shown in Figure 6.4. The Davies–Bouldin Score of 2.39 also suggests a low level of cluster quality for "Classes" compared with that of K-medoids.

Notably, the model achieved the highest silhouette score and Calinski–Harabasz Score among the models, indicating stronger cluster separation and cohesion. In contrast, the BIRCH model obtained the lowest scores, suggesting less distinct clustering. However, all models performed reasonably well in forming clusters and could be considered for categorizing the data.

Both K-medoids and BIRCH have proven to be effective in clustering data from 6G wireless networks. K-medoids outperformed BIRCH, but it is also computationally expensive. BIRCH is less accurate than K-medoids, but it is significantly faster.

6.6 Future Work

Other than K-medoids and BIRCH, several clustering methods have been proposed for 6G wireless communications. These methods include DBSCAN, CURE, and mean shift. The choice of clustering method depends on the specific application and data properties. Future work could investigate other unsupervised learning techniques. Furthermore, with the new era of 6G, clustering methods can be used in new communication scenarios that implement many antennas in multiple dimensions, which are called XL-MIMO systems.

6.7 Conclusion

In the new era of 6G and big data, a method for investigating wireless channel parameters using unsupervised learning methodologies is necessary. Clustering can be used to identify groups of parameters that are used simultaneously in a system. This information can then be used to improve the efficiency and understand the data set. K-medoids and BIRCH are two clustering methods that are effective for 6G wireless communication. The chosen clustering method for 6G wireless communication depends on the specific application. If the accuracy is critical, then K-medoids should be used. If speed is critical, BIRCH should be used. These clustering algorithms can enhance our understanding of propagation channels in the 6G era.

References

[1] Bonaccorso, G. Machine learning algorithms. (Packt Publishing Ltd,2017)

[2] Aldossari, S. & Chen, K. Machine learning for wireless communication channel modeling: An overview. *Wireless Personal Communications*. **106**, pp. 41-70 (2019)

[3] He, R., Li, Q., Ai, B., Geng, Y., Molisch, A., Kristem, V., Zhong, Z. & Yu, J. A kernel-power-density-based algorithm for channel multipath components clustering. *IEEE Transactions on Wireless Communications*. **16**, 7138-7151 (2017)

[4] He, R., Ai, B., Molisch, A., Stuber, G., Li, Q., Zhong, Z. & Yu, J. Clustering enabled wireless channel modeling using big data algorithms. *IEEE Communications Magazine*. **56**, 177-183 (2018)

[5] Czink, N., Tian, R., Wyne, S., Tufvesson, F., Nuutinen, J., Ylitalo, J., Bonek, E. & Molisch, A. Tracking time-variant cluster parameters in MIMO channel measurements. *2007 Second International Conference on Communications and Networking in China*. pp. 1147-1151 (2007)

[6] Spyridis, Y., Lagkas, T., Sarigiannidis, P., Argyriou, V., Sarigiannidis, A., Eleftherakis, G. & Zhang, J. Towards 6G IoT: Tracing mobile sensor nodes with deep learning clustering in UAV networks. *Sensors*. **21**, 3936 (2021)

[7] Dang, S., Amin, O., Shihada, B. & Alouini, M. What should 6G be? *Nature Electronics*. **3**, 20-29 (2020)

[8] Wu, J. Advances in K-means clustering: a data mining thinking. (Springer Science & Business Media, 2012)

[9] Park, H. & Jun, C. Simple and fast algorithms for K-medoids clustering. *Expert Systems With Applications*. **36**, 3336-3341 (2009)

[10] Zhang, T., Ramakrishnan, R. & Livny, M. BIRCH: A new data clustering algorithm and its applications. *Data Mining And Knowledge Discovery*. **1**, pp. 141-182 (1997)

[11] Palacio-Niño, J. & Berzal, F. Evaluation metrics for unsupervised learning algorithms. *ArXiv Preprint ArXiv:1905.05667*. (2019)

[12] Petrovic, S. A comparison between the silhouette index and the Davies–Bouldin index in labelling id clusters. *Proceedings of the 11th Nordic Workshop of Secure IT Systems*. **2006** pp. 53-64 (2006)

[13] Wang, C., You, X., Gao, X., Zhu, X., Li, Z., Zhang, C., Wang, H., Huang, Y., Chen, Y., Haas, H. et al. On the road to 6G: Visions, requirements, key technologies and testbeds. *IEEE Communications Surveys & Tutorials*. (2023)

[14] Ahammed, T., Patgiri, R. & Nayak, S. A vision on the artificial intelligence for 6G communication. *ICT Express*. **9**, 197-210 (2023)

[15] Dogra, A., Jha, R. & Jain, S. A Survey on Beyond 5G Network with the advent of 6G: Architecture and Emerging Technologies. *IEEE Access*. **9**, 67512-67547 (2021)

[16] Krenn, M., Landgraf, J., Foesel, T. & Marquardt, F. Artificial intelligence and machine learning for quantum technologies. *Physical Review A*. **107**, 010101 (2023)

[17] Rappaport, T., MacCartney, G., Samimi, M. & Sun, S. Wideband millimeter-wave propagation measurements and channel models for future wireless communication system design. *IEEE Transactions on Communications*. **63**, 3029-3056 (2015)

[18] May Taniguchi, L. & Abrão, T. Stochastic channel models for massive and extreme large multiple-input multiple-output systems. *Transactions on Emerging Telecommunications Technologies*. **31**, e4099 (2020)

[19] Nuckelt, J., Schack, M. & Kürner, T. Deterministic and stochastic channel models implemented in a physical layer simulator for Car-to-X communications. *Advances in Radio Science*. **9**, 165-171 (2011)

[20] Nawaz, S., Sharma, S., Wyne, S., Patwary, M. & Asaduzzaman, M. Quantum machine learning for 6g communication networks: State-of-the-art and vision for the future. *IEEE Access*. **7**, 46317-46350 (2019)

[21] He, H., Jin, S., Wen, C., Gao, F., Li, G. & Xu, Z. Model-driven deep learning for physical layer communications. *IEEE Wireless Communications*. **26**, 77-83 (2019)

[22] Sarkar, N. & Gul, S. Artificial intelligence-based autonomous UAV networks: A survey. *Drones*. **7**, 322 (2023)

[23] Shyja, V., Ranganathan, G. & Bindhu, V. Link quality and energy efficient optimal simplified cluster based routing scheme to enhance lifetime for wireless body area networks. *Nano Communication Networks* 100465 (2023)

[24] Singh, P., Singh, N., Luxmi, P. & Saxena, A. Artificial intelligence for smart data storage in cloud-based IoT. *Transforming Management With AI, Big-Data, And IoT* 1-15 (2022)

[25] Letaief, K., Shi, Y., Lu, J. & Lu, J. Edge artificial intelligence for 6G: Vision, enabling technologies, and applications. *IEEE Journal on Selected Areas in Communications*. **40**, 5-36 (2021)

[26] Gebremariam, A., Usman, M. & Qaraqe, M. Applications of artificial intelligence and machine learning in the area of SDN and NFV: A survey. *2019 16th International Multi-Conference on Systems, Signals & Devices (SSD)*. pp. 545-549 (2019)

[27] Letaief, K., Chen, W., Shi, Y., Zhang, J. & Zhang, Y. The Roadmap to 6G: AI Empowered Wireless Networks. *IEEE Communications Magazine*. **57**, 84-90 (2019)

[28] Sun, S., MacCartney, G. & Rappaport, T. A novel millimeter-wave channel simulator and applications for 5G wireless communications. *2017 IEEE International Conference On Communications (ICC)*. pp. 1-7 (2017)

[29] Ju, S. & Rappaport, T. Millimeter-wave extended NYUSIM channel model for spatial consistency. *2018 IEEE Global Communications Conference (GLOBECOM)*. pp. 1-6 (2018)

[30] Aldossari, S. & Chen, K. Predicting the path loss of wireless channel models using machine learning techniques in mmwave urban communications. *2019 22nd International Symposium on Wireless Personal Multimedia Communications (WPMC)*. pp. 1-6 (2019)

[31] Seidel, S. & Rappaport, T. 914 MHz path loss prediction models for indoor wireless communications in multifloored buildings. *IEEE Transactions on Antennas and Propagation*. **40**, 207-217 (1992)

[32] Arora, P., Varshney, S. et al. Analysis of k-means and k-medoids algorithm for big data. *Procedia Computer Science*. **78**, 507-512 (2016)

[33] Janaswamy, R. Radiowave propagation and smart antennas for wireless communications. (Springer Science & Business Media,2005)

[34] Zhang, T., Ramakrishnan, R. & Livny, M. BIRCH: an efficient data clustering method for very large databases. *ACM Sigmod Record*. **25**, 103-114 (1996)

7

Network Traffic Data Prediction using Deep Learning

Buminhan Cay[1], Evren Tuna[1], Erdem Demircioglu[1], and Hüseyin Arslan[2]

[1]Department of Research and Development, ULAK Communications Inc., Turkey
[2]School of Engineering and Natural Sciences, Istanbul Medipol University, Turkey
E-mail: buminhan048@gmail.com; evren.tuna@ulakhaberlesme.com.tr; erdem.demircioglu@ulakhaberlesme.com.tr; huseyinarslan@medipol.edu.tr

Abstract

Applying artificial intelligence (AI) and machine learning (ML) is crucial for telecom networks to meet users' diverse service requirements autonomously and robustly. The diversity of services and the increasing device density make this problem even more complex. This chapter explains the importance of AI/ML applications in next-generation mobile networks with focus areas and current literature.

This study presents experimental results where various ML models, trained using user-specific measurements in an urban environment, are employed to classify traffic into six distinct levels. The dataset used in this research is an open-source LTE dataset containing real-time network data shared by Telecom Italia. Besides, it employs supervised and unsupervised learning techniques to classify traffic volume levels in mobile networks. By classifying traffic levels, network slicing is enhanced, and resource allocation is optimized. Ultimately this improves overall network performance.

The results of the study demonstrate the remarkable accuracy achieved by both supervised learning models, deep neural network (DNN) and convolutional neural network (CNN). The DNN model exhibits a steady

DOI: 10.1201/9788743801108-8

increase in accuracy, reaching a peak of 99.5%. Similarly, the CNN model achieves an impressive accuracy of 99%. These findings underscore the reliability and similarity between the two models despite the DNN model's higher parameter count. From a computational standpoint, the CNN model is more efficient regarding training load, offering a compelling alternative for mobile traffic prediction. In unsupervised learning, this study reveals promising results by integrating K-means clustering and principal component analysis (PCA) and combining PCA with K-means clustering, which results in a 32.76% performance improvement in unsupervised learning. Classifying different services based on user-specific measurements is critical for triggering network configurations specific to different network slices.

Keywords: Mobile networks, service classification, neural network, cognitive networks

7.1 Introduction

The initiation of 5G and beyond technologies establishes new complications for mobile service providers, and incorporating AI methods into their networks is a solid approach to handling these difficulties. The ability to apply AI techniques is a major research area for future-generation telecom networks [1–3].

AI is now being combined into mobile operator networks, with a principal concentration on lowering capital expenses, improving network performance, and developing additional income flows. In addition, AI will be critical for advancing customer service and experience. Finally, implementing AI into network operations establishes novel data challenges, although it answers system complexities [4].

Core sections, in which the biggest potential for AI investments returns, are listed below [4]:

- Network planning
- Network performance management
- Service level agreement (SLA) management
- Product lifecycle management
- Network operations and management
- Revenue management

Figure 7.1 depicts the generic applications of 5G and beyond communication technologies, forming a triangle of enhanced mobile broadband (eMBB), ultrareliable and low latency communication (URLLC), and massive machine-type communication (mMTC). eMBB addresses the need for increased spectrum bands to support high-definition videos and online gaming, while URLLC ensures high-speed responsiveness for applications such as vehicle-to-everything (V2X) and autonomous transportation. Meanwhile, mMTC facilitates wide sensor communication between machines and human interfaces. With the advent of 6G networks, a hexagonal expansion of communication scenarios emerges, including immersive communication, massive communication, and hyper reliable and low-latency communication, complementing the foundational pillars of eMBB, URLLC, and mMTC. Additionally, 6G incorporates integrated sensing and communication, AI and communication, and ubiquitous connectivity as pivotal technologies, emphasizing the evolution toward more comprehensive and intelligent communication ecosystems. Among these application scenarios, three technologies are becoming increasingly important, and AI/ML can be shown as the most distinguishing technology for 6G [5–7].

Since the data load for operator networks is increasing, novel methods and tools are essential to handle these data loads. Besides, AI tools are essential for generating cost-efficient solutions. The tremendous evolution of neural network (NN) over the last decade has enabled researchers to implement their solutions to complicated problems [8].

ML is extensively applied across a wide range of domains, including communication and networking. This technology, anticipated to gain even more significance, aims to achieve heightened levels of interdevice communication, optimize network performance, and improve the efficiency of cellular networks [9].

As ML becomes increasingly integrated within the realm of 5G and gains prominence within the evolving landscape of 6G, its potential expands beyond mere efficiency, encompassing many benefits across various sectors. A pivotal aspect of leveraging NNs is its capability to establish intricate connections within complex data sets. The ease of establishing these connections not only facilitates efficient data classification and forecasting but also contributes to time and energy savings [10].

Figure 7.2 illustrates the aspects of AI/ML improving the 6G network performance management. As AI tools are used to manage mobility and user behavior for localization and traffic, better network performance can be generated as an output. In addition, network efficiency can be enhanced by resource utilization, admission, and congestion control. In terms of network security, analyzing massive data can enforce threat detection and better user

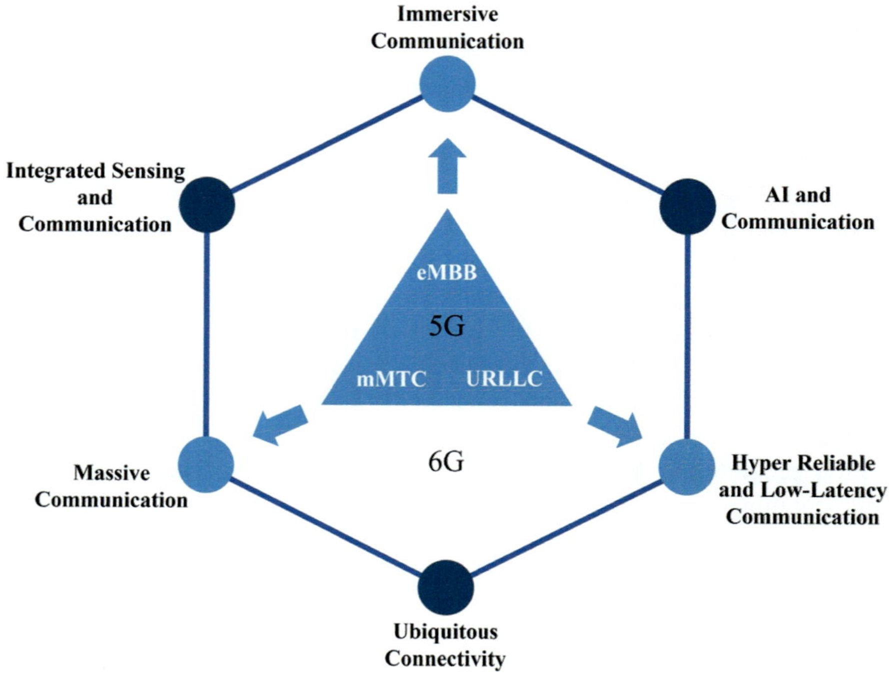

Figure 7.1 6G use cases and technologies.

behavior localization. Finally, small cells and network topologies can be used for better 6G deployment [4].

Deep learning (DL) offers an impressive ability to discover and learn patterns among data using multilayered NNs. The application of DL on large data sets has achieved remarkable success in many areas, such as image recognition, natural language processing, gaming, and autonomous driving. With its flexible structure and broad application areas, DL constitutes one of the most crucial artificial intelligence research areas.

While internet, call, and short message service (SMS) remain fundamental pillars of mobile networks, as more complex services are delved into, such as extended reality and remote control, it is easy to assume that the requirements for these services may not be as demanding. However, the dynamic nature of modern society, which includes unexpected events like natural disasters or sudden surges in crowd movement in any location, underscores the critical need for ML-driven predictions.

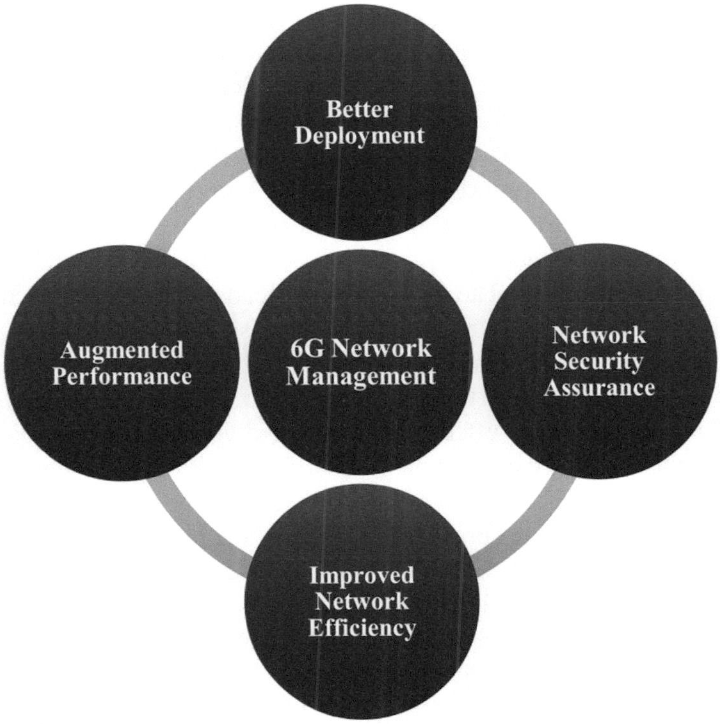

Figure 7.2 6G network performance management via AI/ML tools [4].

Network traffic prediction has received considerable attention. Some studies address this challenge using regression techniques to forecast continuous values, such as traffic volume. In contrast, others utilize classification-based approaches to categorize traffic into distinct classes or levels. In the domain of regression-based traffic prediction, notable works include [11–14], which employ long short-term memory (LSTM) networks to forecast traffic in mobile networks. Additionally, [15] utilizes a gated recurrent unit (GRU) for traffic prediction in similar contexts. The studies [16, 17] leverage CNNs for traffic prediction in mobile networks, while [18] combines LSTM networks with CNNs.

Furthermore, [19] explores DNN architectures for predicting future traffic demands in communication networks. In [20], LSTM, GRU, and online sequential extreme learning machine (OS-ELM) were used to predict network traffic data and compared to these different techniques for determining the best network regarding prediction accuracy and computational time. LSTM and GRU models, while effective in capturing temporal dependencies, often suffer from slower training times and higher computational loads due to their recurrent nature and complex architectures. In contrast, CNNs offer a more efficient solution for specific applications, requiring fewer parameters and offering faster inference times without sacrificing predictive performance. As classification-based approaches are explored, the distinctiveness of regression-based studies in predicting continuous values such as traffic volume is worth noting. The shift to classification introduces a change in the problem's nature, rendering direct comparisons between regression and classification methodologies challenging.

The classification of services, tailored to different user device types and capabilities, is crucial to effectively meeting diverse service requirements. This multifaceted landscape not only underlines the ever-evolving challenges within mobile network services but also presents numerous opportunities for innovation. The use of accurate DL-based models to classify these services and make accurate predictions [21] is a key area of research. There are numerous studies on classification-based approaches. In [22], a high-level overview of these techniques is presented, along with a discussion of their applications in local and wide area networks. Works such as [23, 24] delve into the classification of network traffic based on application types using CNNs.

Moreover, to determine their suitability for specific applications, they explore the applicability of various ML and statistical models, including NNs, linear regression, support vector machines, PCA, statistical models, and linear time series models. Another critical issue in DL is data management. Data set collection and preprocessing stages are critical steps for efficient training and validation. It should be noted that the statistical analysis of the data set needs to be done carefully to divide the data set into train, validation, and test sets.

This study uses supervised and unsupervised learning techniques to classify different network types and benchmark the performance of techniques such as CNN, DNN, and K-means clustering. This chapter is structured as follows: Section 7.2 presents a comprehensive description of the utilized data set. Section 7.3 provides the architecture of different models with their hyperparameters. Section 7.4 benchmarks the performance of different models. Finally, Section 7.5 concludes the book chapter.

7.2 Data Set Description and Framework Architecture

In this study, a data set from a real and live mobile network operated by Telecom Italia is utilized. The data set comprises the call detail records (CDRs) of the cellular network within the urban area. These CDRs serve the dual purpose of tracking user behavior for invoicing and network management. The data set includes the user's service-based traffic volumes of SMS-in, SMS-out, call-in, call-out, and internet activities. The data set contains information on network traffic for six months. Table 7.1 illustrates the features of the data set. For more detailed information about these features, please refer to [25].

This study involved a thorough preprocessing of the data set. Rows with null elements are meticulously removed, ensuring that the statistics of the feature set remain unaffected. This step was crucial to guarantee that only SMS-in, SMS-out, call-in, call-out, and internet activities were stored, eliminating any potential noise in the data. After these processes, a new feature called "total activity" was generated, which consists of the SMS-in, SMS-out, call-in, call-out, and internet traffic logs. Furthermore, the data set was divided into three parts – training, validation, and testing to ensure the robustness of this analysis.

In the context of 5G and beyond networks, measuring and accurately predicting traffic volume across various services remains unchanged. While the service types provided over the LTE network in this study might initially appear outdated and their relevance to 5G and beyond could be questioned, the focus on traffic volume data remains central to addressing these challenges. The data consists of traffic volume values for these services, and despite potential network architectures and technology shifts, the fundamental need to monitor and manage traffic volume for different services remains paramount. This research is dedicated to unraveling and fine-tuning

Table 7.1 Data set description.

Name	Description
Square ID	Represents a part of the grid
Start time	Start time of the period
Country code	Represents the country code
SMS-in	Incoming SMS service-based traffic volume within the square ID
SMS-out	Outgoing SMS service-based traffic volume within the square ID
Call-in	Incoming call service-based traffic volume within the square ID
Call-out	Outgoing call service-based traffic volume within the square ID
Internet	Internet service-based traffic volume within the square ID

the intricacies of traffic volume dynamics, which is critical for optimizing the performance of upcoming mobile communication infrastructures.

When we look at the studies in the literature, we see that network traffic prediction studies use data sets obtained either in a simulation environment or from real and live cellular networks. Although access and data collection are difficult due to privacy concerns, data sets obtained from real and live networks are more useful for developing ML-based solutions that consider instantaneous environmental changes challenging mobile networks (e.g., spatial-temporal effects and sudden crowd movements). Therefore, a publicly available data set from a real and live network has been used in this study. According to the 3GPP in [26–28], it is seen that the traffic volume metric is a defined metric for Release 19 and beyond. This study aims to propose an ML-based framework that will achieve cellular traffic volume prediction under different environmental conditions and service dynamics to meet the desired requirements. A new data set collected from a 5G and beyond communication network will have different patterns regarding the services it will enable and the associated environmental conditions. Therefore, a new data set obtained from the next-generation network will be more than just a different patterned data for this solution. These changes will require adjustments in the ML model architecture and its related hyperparameters.

The framework used in this study encompasses several essential functions crucial for the implementation of models, as detailed in Figure 7.3. It begins with the data collection function, which provides input data essential for

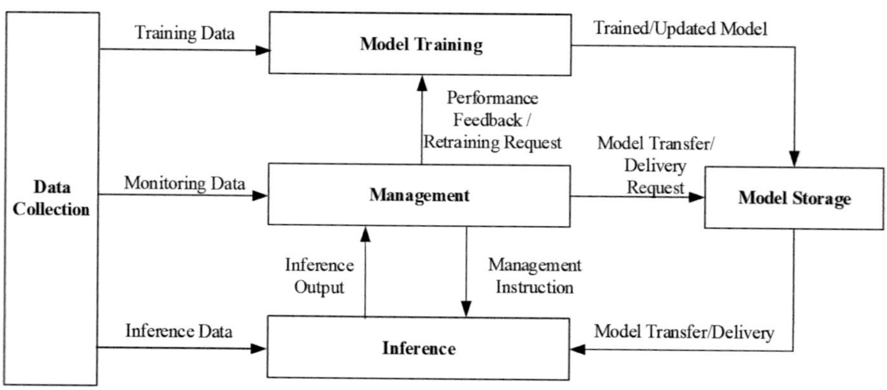

Figure 7.3 Framework architecture [29].

subsequent functions such as model training, model management, and model inference. On the other hand, the model training function focuses on training, validating, and testing models, emphasizing data preparation and cleaning derived from training data. Following this, the management function oversees the operation and monitoring of models, including decision-making based on data received from data collection and inference functions. The inference function, meanwhile, facilitates the application of models to produce outputs, utilizing data from the data collection function. Finally, the model storage function plays a pivotal role in storing trained or updated models for deployment in the inference function, ensuring the seamless operation of the overall framework.

7.3 Model Architectures

This section presents different ML models which employ both supervised and unsupervised learning.

7.3.1 DNN

A DNN-based model was chosen for its technical capabilities and versatility. DNNs excel in capturing complex patterns and relationships within various types of data, thanks to their ability to model nonlinear interactions across multiple layers of abstraction. This adaptability is a key strength, as DNN architecture can be generalized across diverse data sets, including structured and unstructured data. DNNs have shown remarkable success in tasks where feature extraction and high-dimensional data processing are crucial. The DNN model architecture is illustrated in Figure 7.4. The depth of the model, which includes six hidden layers, was determined through rigorous hyperparameter tuning to ensure optimal performance. Each hidden layer employs the ReLU activation function and consists of 200 hidden units to balance complexity and computational efficiency.

The hyperparameter tuning process also determined the training parameters, including 30 epochs and a batch size of 10^4. These values provided sufficient learning iterations while maintaining manageable training times. The Adam optimizer was selected for its adaptive learning rate capabilities, which enhance convergence speed and stability. Categorical cross-entropy was used as the loss function to handle the multiclass classification problem presented by the data set effectively. The parameter number of proposed DNN model is 243606.

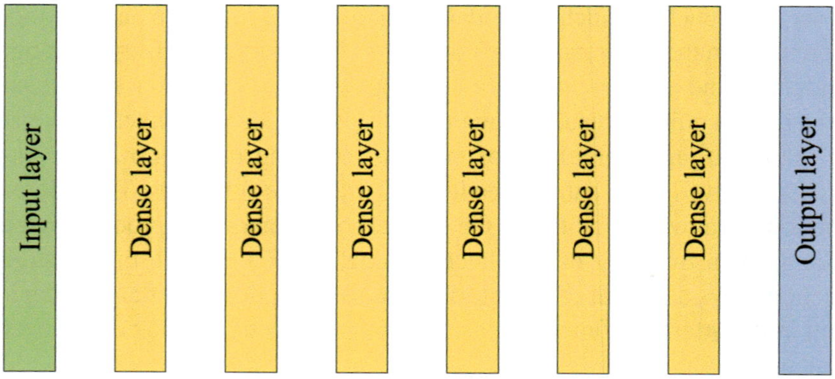

Figure 7.4 DNN model architecture.

7.3.2 CNN

A CNN model is developed to investigate its performance and effectiveness compared to the DNN baseline. CNNs are designed to handle grid-like data structures, such as images and time-series data, by automatically and adaptively learning spatial hierarchies of features through convolutional layers. This architecture is particularly effective in capturing intricate features and patterns within the data, making CNNs a powerful tool for tasks requiring spatial understanding.

The CNN architecture comprises multiple layers to efficiently extract and learn these spatial hierarchies. As shown in Figure 7.5, the model consists of five hidden layers. The architecture begins with two convolutional layers, which apply filters to the input data, capturing local patterns and features. Each convolutional layer is followed by an activation function (ReLU) to introduce nonlinearity and enhance the model's ability to learn complex patterns.

Following the convolutional layers, a flattened layer converts the 2D feature maps output by the convolutional layers into a 1D vector, preparing the data for the fully connected layers. The model then includes two dense layers; the first dense layer further processes the flattened input, and the final dense layer outputs the class probabilities using the SoftMax activation function.

For this model, the hyperparameters are precisely tuned to optimize performance. The epoch is set to 30 and the batch size to 10^4, ensuring ample training iterations for convergence. Categorical cross-entropy is employed as the loss function, a fitting choice for the multiclass classification problem,

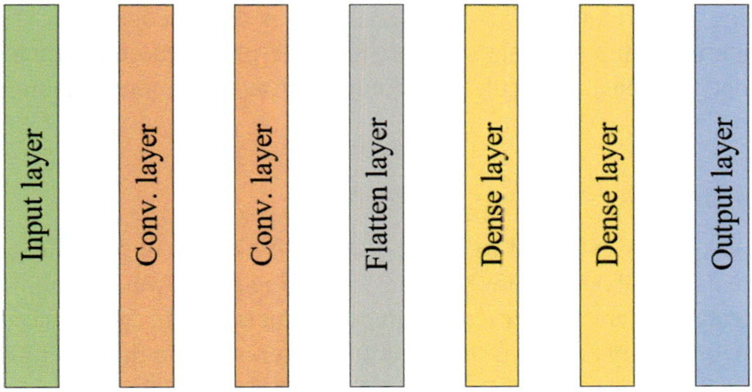

Figure 7.5 CNN model architecture.

and utilized the Adam optimizer for its adaptive learning rate and efficiency in handling sparse gradients. The number of parameters in this model is 15430, a testament to the commitment to striking a balance between model complexity and computational feasibility.

7.3.3 K-means clustering

In this study, following an initial exploration using supervised learning models, the investigation's focus shifts to the unsupervised learning domain. The primary aim is to uncover latent structures within the data without relying on pre-existing labels. To achieve this, the robust unsupervised machine learning technique K-means clustering is employed, which groups data points based on similarity.

K-means clustering operates by partitioning the data set into K clusters, where each data point belongs to the cluster with the nearest mean, serving as a prototype of the cluster. This iterative process continues until convergence, effectively minimizing the within-cluster variance. Through K-means clustering, it aims to reveal hidden patterns, relationships, and inherent groupings within the data set that may not be apparent through supervised learning approaches.

When applying K-means clustering, a deliberate choice is made to set the number of clusters, K, to 6. This decision was guided by an analysis of the data set's characteristics and aimed to strike a balance between granularity and interpretability. The K-means algorithm traverses the data set, adjusting cluster centroids and reassigning data points to clusters to optimize the overall clustering performance. This approach aids in the discovery of structures and patterns within the data set, enriching the proposed comprehension of the underlying data distribution and contributing to a more comprehensive analysis [30].

7.3.4 PCA

This approach is extended to incorporate PCA with K-means clustering to enhance data analysis. While K-means clustering is a powerful technique for uncovering inherent data structures, PCA complements this process by reducing dimensionality and capturing essential features within the data [30]. PCA is a dimensionality reduction technique that transforms high-dimensional data into a lower-dimensional space while preserving the essential variance in the data. By retaining the most informative dimensions, PCA enables a more concise data set representation, facilitating improved clustering performance.

The integration of PCA with K-means clustering offers a promising potential to not only identify data clusters but also enhance the quality of the information used for clustering. By reducing the dimensionality of the data through PCA, it is aimed to mitigate the curse of dimensionality and improve the efficiency and effectiveness of the clustering process. This combined approach allows for a more focused analysis of the most relevant features, leading to more meaningful cluster assignments and opening exciting possibilities for data analysis.

7.4 Results

This section provides the performance results of network traffic classification using different ML models. Initially, the change in the accuracy of DNN and CNN models for supervised learning is demonstrated. Then, the performance results of K-means clustering and K-means clustering with PCA for unsupervised learning are presented.

The DNN and CNN models have produced notably similar accuracy results for supervised learning. As illustrated in Figure 7.6, the DNN model's accuracy is 95% and steadily ascends to a peak of 99.5%. These exceptional training outcomes are substantiated by test results, demonstrating an impressive accuracy of 99.5%. Figure 7.7 depicts the performance results of the CNN model. Accuracy reaches an impressive 99%. These findings are highly accurate, and the test results confirm an accuracy level of 99%. When comparing these results with those of the DNN model, they exhibit

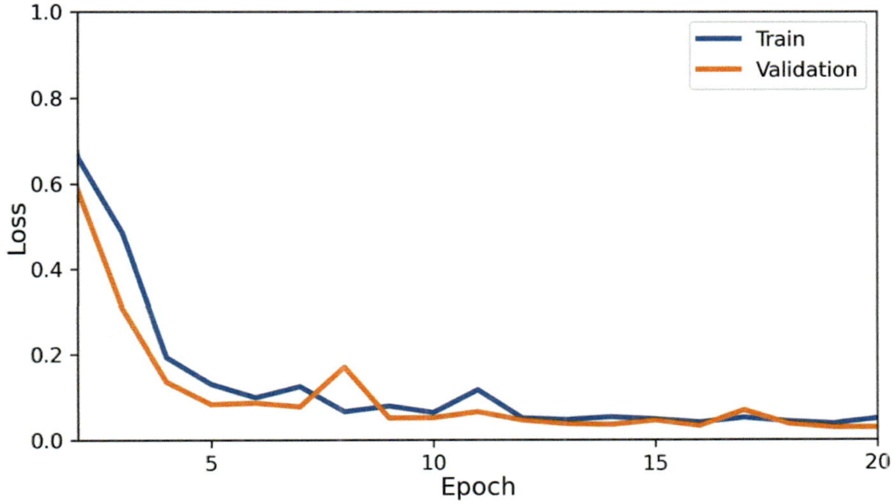

Figure 7.6 Train and validation loss for DNN model.

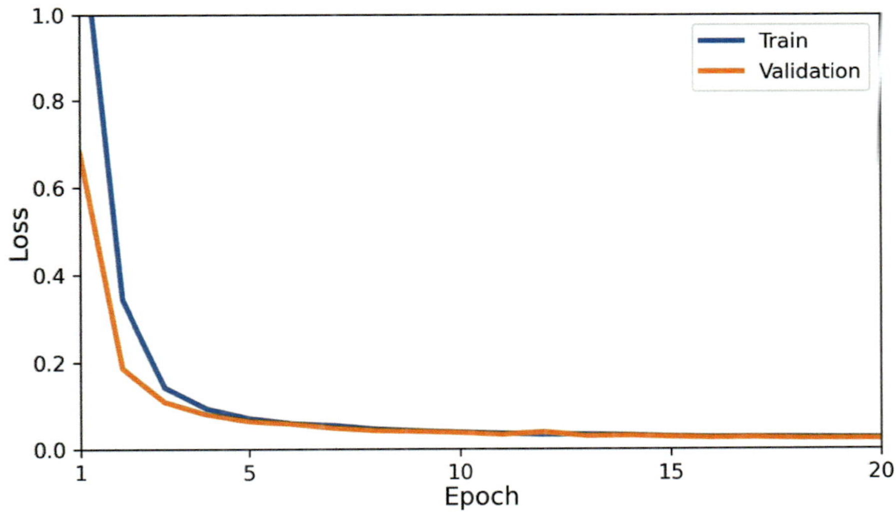

Figure 7.7 Train and validation loss for CNN model.

remarkable similarity and reliability. Furthermore, the DNN model boasts approximately 15 times the number of parameters compared to the CNN model. From the perspective of model complexity, the utilization of the CNN model proves to be more computationally efficient in terms of training load.

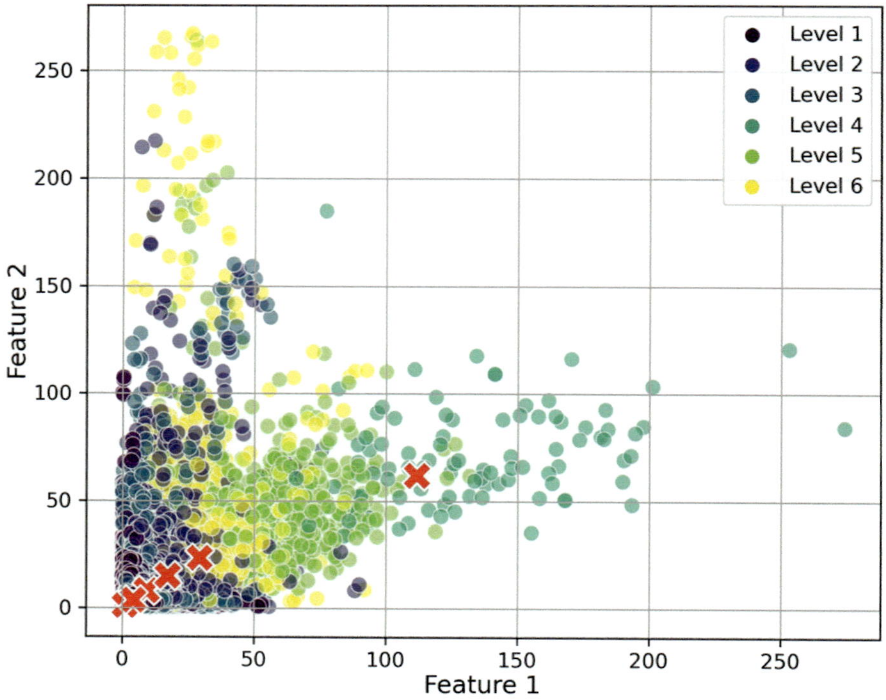

Figure 7.8 K-means clustering.

For unsupervised learning, it is observed that the combination of K-means clustering and PCA significantly improves performance. Figure 7.8 depicts the performance results of K-means clustering. The test result is only 58%. While a decrease in accuracy compared to supervised learning was expected, it is observed that the data set's existing dimensionality significantly impairs performance [30].

Hence, to mitigate this issue, PCA is applied to reduce the dimensionality of the additional feature set. Figure 7.9 shows that the test result is 77%. This model has more accurate results than the K-means clustering model but still needs to catch up to the DNN and CNN models. In a mobile network, there may be a need to employ unsupervised learning under certain conditions. In such cases, operators or vendors must determine performance thresholds carefully. This is essential to select the most suitable model that achieves the desired performance rate while optimizing model complexity and computational load.

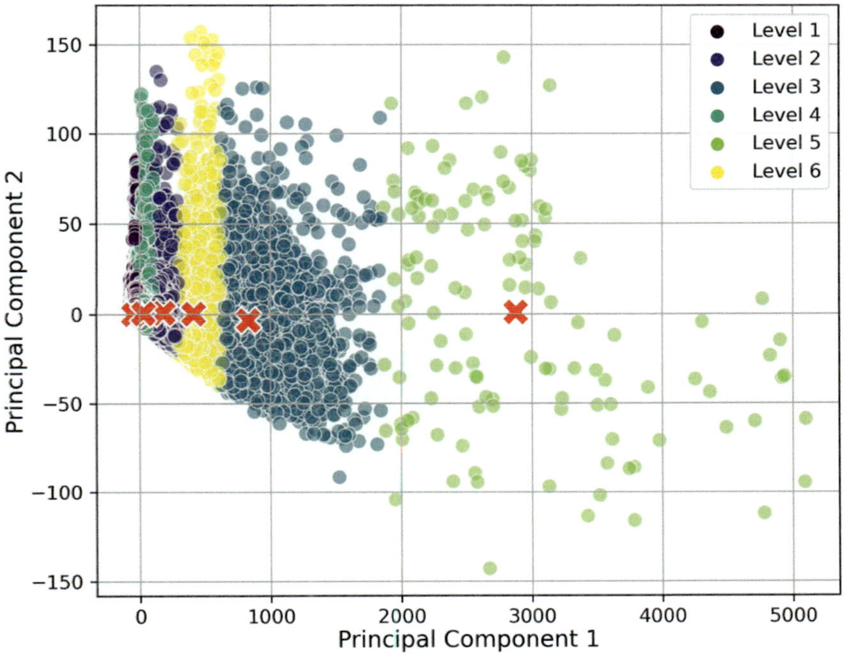

Figure 7.9 K-means clustering with PCA.

7.5 Conclusion

Due to the increasing device density and the diversity of network traffic types in telecom networks, it is essential to classify the different traffic levels to make networks more dynamic and robust. In this study, we benchmarked and evaluated the classification results of different models for supervised and unsupervised learning scenarios. For supervised learning, the DNN model, which has approximately one-fifteenth the number of parameters as the CNN model, has yielded performance results similar to that of the DNN model. Consequently, regarding model complexity, the CNN model's ability to learn spatial hierarchies of the data has improved performance. Additionally, using PCA in conjunction with K-means clustering has resulted in a 32.76% performance enhancement for unsupervised learning. The unsupervised models are inefficient compared to the supervised models for traffic prediction applications. On the other hand, utilizing well-defined thresholds for performance, model complexity, and computational load is essential when designing a model specific to traffic classification in mobile networks. The classification of different types of traffic in a mobile network and triggering optimization actions based on traffic type will be critical in future networks. Some studies for further exploration exist in future work. Delving deeper into hybrid approaches that amalgamate the strengths of supervised and unsupervised learning methods could yield more robust and versatile traffic classification models. Moreover, investigating the feasibility and efficacy of deploying these models in real-world network environments and assessing their scalability and interoperability will be crucial.

References

[1] W. Saad, M. Bennis, and M. Chen, "A vision of 6G wireless systems: Applications, trends, technologies, and open research problems," IEEE Networks., vol. 34, no. 3, pp. 134-142, May 2020.

[2] F. Tariq, M. R. A. Khandaker, K.-K. Wong, M. A. Imran, M. Bennis, and M. Debbah, "A speculative study on 6G," IEEE Wireless Communications, vol. 27, no. 4, pp. 118-125, Aug. 2020.

[3] M. Z. Chowdhury, M. Shahjalal, S. Ahmed, and Y. M. Jang, "6G wireless communication systems: Applications, requirements, technologies, challenges, and research directions," IEEE Open J. Communications Society, vol. 1, pp. 957-975, 2020.

[4] Ericsson Technical Report "What role will 5G and AI have in the mobile networks of the future?" Available at: https://www.ericsson.com/en/network-services/ai-5g-networks.

[5] B. Li, Z. Fei, and Y. Zhang, "UAV communications for 5G and beyond: Recent advances and future trends", IEEE Internet Things Journal, vol. 6, no. 2, pp. 2241- 2263, Apr. 2019.

[6] S. Chen, Y.-C. Liang, S. Sun, S. Kang, W. Cheng, and M. Peng, "Vision, requirements, and technology trend of 6G: How to tackle the challenges of system coverage, capacity, user data-rate and movement speed," IEEE Wireless Communication, vol. 27, no. 2, pp. 218-228, Apr. 2020.

[7] Y. Yuan, Y. Zhao, B. Zong, and S. Parolari, "Potential key technologies for 6G mobile communications," Sci. China Inf. Sci., vol. 63, pp. 1-19, May 2020.

[8] J. Kaur, M. A. Khan, M. Iftikhar, M. Imran and Q. Emad Ul Haq, "Machine Learning Techniques for 5G and Beyond," in IEEE Access, vol. 9, pp. 23472-23488, 2021.

[9] C. Jiang et al., "Machine learning paradigms for next-generation wireless networks," IEEE Wireless Communication, vol. 24, no. 2, pp. 98–105, Apr. 2017.

[10] K. Aykurt and W. Kellerer, "Autonomous Network Management in Multi-Domain 6G Networks based on Graph Neural Networks," 2023 IEEE 9th International Conference on Network Softwarization (NetSoft), Madrid, Spain, 2023, pp. 338-341.

[11] J. Feng, X. Chen, R. Gao, M. Zeng and Y. Li, "DeepTP: An End-to-End Neural Network for Mobile Cellular Traffic Prediction," in IEEE Network, vol. 32, no. 6, pp. 108-115, November/December 2018.

[12] I. Alawe, A. Ksentini, Y. Hadjadj-Aoul and P. Bertin, "Improving Traffic Forecasting for 5G Core Network Scalability: A Machine Learning Approach," in IEEE Network, vol. 32, no. 6, pp. 42-49, November/December 2018.

[13] E. Tuna and A. Soysal, "Multivariate and Multi-step Traffic Prediction for NextG Networks with SLA Violation Constraints," 2023 International Balkan Conference on Communications and Networking (BalkanCom), İstanbul, Turkiye, 2023, pp. 1-5.

[14] J. Wang et al., "Spatiotemporal modeling and prediction in cellular networks: A big data enabled deep learning approach," IEEE INFOCOM 2017 - IEEE Conference on Computer Communications, Atlanta, GA, USA, 2017, pp. 1-9.

[15] H. Chergui and C. Verikoukis, "Offline SLA-Constrained Deep Learning for 5G Networks Reliable and Dynamic End-to-End Slicing," in IEEE Journal on Selected Areas in Communications, vol. 38, no. 2, pp. 350-360, Feb. 2020.

[16] C. Zhang, H. Zhang, D. Yuan and M. Zhang, "Citywide Cellular Traffic Prediction Based on Densely Connected Convolutional Neural Networks," in IEEE Communications Letters, vol. 22, no. 8, pp. 1656-1659, Aug. 2018.

[17] D. Bega, M. Gramaglia, M. Fiore, A. Banchs and X. Costa-Pérez, "DeepCog: Optimizing Resource Provisioning in Network Slicing With AI-Based Capacity Forecasting," in IEEE Journal on Selected Areas in Communications, vol. 38, no. 2, pp. 361-376, Feb. 2020.

[18] S. Zhao et al., "Cellular Network Traffic Prediction Incorporating Handover: A Graph Convolutional Approach," 2020 17th Annual IEEE International Conference on Sensing, Communication, and Networking (SECON), Como, Italy, 2020, pp. 1-9.

[19] A. Thantharate, R. Paropkari, V. Walunj and C. Beard, "DeepSlice: A Deep Learning Approach towards an Efficient and Reliable Network Slicing in 5G Networks," 2019 IEEE 10th Annual Ubiquitous Computing, Electronics \& Mobile Communication Conference (UEMCON), New York, NY, USA, 2019, pp. 0762-0767.

[20] F. Rau, I. Soto, D. Zabala-Blanco, C. Azurdia-Meza, M. Ijaz, S. Ekpo, and S. Gutierrez, "A novel traffic prediction method using machine learning for energy efficiency in service provider networks," Sensors, vol. 23, no. 11, 2023.

[21] P. Kaushik, S. S. Singh, and P. K. Yadav, "Traffic prediction in telecom systems using deep learning," 2018 7th International Conference on Reliability, Infocom Technologies and Optimization (Trends and Future Directions) (ICRITO), pp. 302–307, 2018.

[22] A. Chen, J. Law, and M. Aibin, "A survey on traffic prediction techniques using artificial intelligence for communication networks," Telecom, vol. 2, no. 4, pp. 518–535, 2021.

[23] R. Toncharoen and M. Piantanakulchai, "Traffic State Prediction Using Convolutional Neural Network," 2018 15th International Joint Conference on Computer Science and Software Engineering (JCSSE), Nakhonpathom, Thailand, 2018, pp. 1-6.

[24] T. Ko, S. M. Raza, D. T. Binh, M. Kim and H. Choo, "Network Prediction with Traffic Gradient Classification using Convolutional Neural Networks," 2020 14th International Conference on Ubiquitous Information Management and Communication (IMCOM), Taichung, Taiwan, 2020, pp. 1-4.

[25] C.-W. Huang, C.-T. Chiang, and Q. Li, "A study of deep learning networks on mobile traffic forecasting," in 2017 IEEE 28th Annual International Symposium on Personal, Indoor, and Mobile Radio Communications (PIMRC), 2017, pp. 1–6.

[26] "Telecommunication management; Performance Management (PM); Performance measurements E-UTRAN," 3GPP, Technical Specification (TS) 32.425, 06 2021, version 17.1.0.

[27] "Telecommunication management; Performance Management (PM); Concept and requirements," 3GPP, Technical Specification (TS) 32.401, 04 2022, version 17.0.0.

[28] "Management and orchestration; 5G performance measurements Telecommunication management," 3GPP, Technical Specification (TS) 28.552, 06 2024, version 19.0.0.

[29] "Study on Artificial Intelligence (AI)/Machine Learning (ML) for NR air interface," 3GPP, Technical Report (TR) 38.843 V18.0.0.

[30] E. Alpaydin, Introduction to Machine Learning, 2nd ed. The MIT Press, 2010.

8

Area Sweep Coverage: A Trade-off between Tour Length and Memory Requirements

Rinku Sen, Saumya Jaipuria, and Rajib K. Das

Department of Computer Science and Engineering, University of Calcutta, India
E-mail: rineelsen@gmail.com; saumyajaipuria_r@isical.ac.in; rajib.k.das@ieee.org

Abstract

Area sweep coverage is one of the essential applications of wireless sensor networks. Prior research indicated that finding an optimal schedule to sweep a particular region is NP-complete, and many researchers attempted to find an approximate solution. One approach based on dividing the surveillance region into squares of side $\sqrt{2}r_s$ (r_s is the sensing radius of the mobile device sweeping the area) gives a tour of length at most $\sqrt{2}$ times the optimal. This work shows that an approximation ratio of $4/3$ on tour length is possible by dividing the region into regular hexagons of side length r_s. Minimizing the tour length is critical in reducing the energy required by mobile devices and the number of mobile devices. We have improved the approximation ratio further at the cost of an increase in the memory requirement of mobile devices. We have simulated our algorithm by varying the dimensions of the surveillance region and the sensing radius r_s. Experimental results show that an approximation ratio close to 1.2 is achievable while keeping the memory requirement within a factor 2 of the optimal.

Keywords: Mobile sensors, wireless sensor networks, sweep coverage, Hamiltonian tour

8.1 Introduction

Wireless sensor network is a self-configured network that interacts with and monitors various environmental conditions. The sensors gather the required

DOI: 10.1201/9788743801108-9

information and pass it to the (base station) sink that interprets the data and initiates necessary action. So, to serve the purpose of the network, we must deploy sufficient active sensors to cover the entire surveillance region. One measures the effectiveness of coverage in wireless sensor networks based on some criteria. Depending on the processing requirement of the applications, one may have three basic types of coverage – area coverage, barrier coverage, and target coverage. For *area coverage*, one requires the sensors to cover the entire area (or a high percentage) of the surveillance region. A random deployment to achieve this may require too many sensors and could be less cost-efficient. So, the primary focus on area coverage has been to ensure coverage with the minimum number of sensors. From a practical aspect, area coverage is difficult to solve as it involves converting a continuous region into a discrete space [9]. There are two approaches to solving this problem. One can deploy static nodes with predetermined locations and thus ensure coverage with the minimum cost of sensors. However, this approach fails when the deployment of mobile sensors depends on the events occurring in the environment. The other method is to deploy mobile sensors and allow them to change their positions by exchanging location information with neighbors. The works in [1] and [10] have tried to address this problem using probabilistic coverage models.

In some applications, continuous monitoring of the target area is not necessary. Because sensors send data to a base station (BS) only at regular intervals and not always, it is sufficient to plan the movements of the mobile sensors to ensure that every point within the area comes within the sensing range of at least one mobile device regularly. This type of coverage is known as *sweep coverage*, and it is a highly efficient way to monitor an area. When planning the trajectory of mobile sensors for sweep coverage, it is crucial to ensure that the interval between successive visits of a point remains within a given value (known as the *sweep period*). For a reasonable range of the speed of the mobile device (say a few meters per second) and the sweep period (around 15 to 20 seconds), the number of mobile devices required to maintain sweep coverage is significantly less than that with static sensors (even with a deterministic deployment). When solving the sweep coverage problem, the factors that need optimization are the total movement of a mobile device to complete a tour (starting from BS, roaming the target area, and returning to BS) and the number of mobile devices employed. To minimize the length of such a tour, we must consider a single tour traversing the target area such that every point in it is within a distance of r_s from the tour path. In this case, if the sensing radius of the mobile device is r_s, an optimal tour must be of length at least $A/2r_\mathrm{s}$ for an area A [3]. The work in [3] shows that for their proposed algorithms, the tour lengths approximate $\sqrt{2}$ times the optimal. To achieve this, they divide a rectangular area into squares whose diagonals are of length $2r_\mathrm{s}$, and the tour passes through the center of all these squares.

Our approach is to find a tour traversing a set of points in rows. By keeping the horizontal spacing between the points equal to $\sqrt{3}r_\mathrm{s}$ and the gap between rows $3r_\mathrm{s}/2$, the tour length can be made approximately equal to $4/3$ times the optimal, thereby giving a better approximation ratio. We can further improve the approximation ratio by reducing the horizontal spacing between the points and increasing the gap between the rows. This method requires mobile devices to store more sensed data. In our algorithm, we can specify how much extra memory we want to provide to the mobile device and plan the tour while keeping the memory requirements within that limit. More sensed data will also cause the mobile devices to transmit more data to the sink at the end of the tour. However, this will not be an issue if the wireless transmission uses 5G/6G technology that can provide higher bandwidth and data rates. We have run our algorithms by varying the dimensions of the target area and the sensing radius. Our results clearly show the trade-off between the tour length and memory requirements.

The rest of the chapter goes as follows. The other related works appear in the next section. Section 8.3 presents the underlying assumptions and the problem scenario and formally defines the problem. Section 8.4 describes and analyzes the proposed algorithm. Section 8.5 presents the experimental results, followed by the conclusion in Section 8.6.

8.2 Related Works

Area coverage is an essential field of study that aims to deploy sensor nodes in a way that covers the entire area. This problem is similar to the Art Gallery Problem, where we need to monitor an environment by placing a minimum number of cameras. To achieve total coverage, [8] suggests deploying static sensors at predetermined locations. However, this strategy may not be suitable for certain situations, such as underwater or volcanic areas, where one must deploy sensors randomly. Unfortunately, this often results in an uneven distribution of sensors, leading to some subregions being over or undercovered. To address this issue, sensor nodes must relocate to improve coverage using potential induced fields [4] or virtual force [6] among the sensor nodes.

In other applications, it may be sufficient to visit all points in the region at regular intervals using mobile devices. The *sweep coverage* problem proposed in [5] aims to find the minimum number of mobile devices required to cover a region within a given sweep period, given the speed of the mobile devices.

Many researches have proposed approximation and heuristic algorithms to solve the sweep coverage problem. They mainly focus on covering a set of target locations within the area with the minimum number of mobile sensor nodes in the shortest possible time or distance. However, the algorithms for target sweep coverage may not always be suitable for area sweep coverage because it is difficult to determine the strategy for covering uncountable points. Therefore, researchers like [2] assume that mobile sensor nodes are discs of radius r. They divide the area into square grids of side $\sqrt{2}r$, use the center of each square as the position to place a sensor, and then apply the two-approximation point coverage algorithm to determine the number of mobile sensor nodes required to cover these points. They found the area coverage approximation ratio to be $\mathbf{2\sqrt{2}}$. They also extended their problem to the rectangular region in [3] and assumed that the sides of the rectangle are divisible by $\sqrt{2}r$. The approximation ratio for their algorithm is $\sqrt{2}$ for even-sized grids and $\left(\sqrt{2} + \frac{4-2\sqrt{2}}{A/r^2}\right)$, where A is the area of the rectangle, for odd-sized grids. In [7], the area is divided into grids so that the diagonal of each subarea is less than or equal to the communication radius of the mobile sensor. They use their point sweep coverage algorithm (P-SCA) within each subarea to ensure coverage of all the points within that subregion. Their proposed area sweep coverage algorithm (A-SCA) has an approximation ratio of $1.5\sqrt{2}$.

8.3 System Model and Problem Definition

We address the area sweep coverage problem based on the following assumptions:

(a) The surveillance region R is rectangular of area A of height H and width W.
(b) A set of mobile sensors roam around the region to monitor the area R.
(c) A point p in the area is under the cover of a sensor s if it lies within a circle of radius r_s (sensing radius) of s.
(d) The mobile sensors can move v meters/second.
(e) A static sink or base station (BS) is within the surveillance region, and the mobile sensors deliver the sensed data at the sink.

Our objective is to design a plan of movement for the mobile sensors satisfying the following conditions:

1. Every point within R should come within the sensing range of at least one mobile sensor at periodic intervals of period T called the sweep period.
2. The sensors have limited memory capacity and can store the sensed data about a limited number of points.

Now we formally state the Area Sweep Coverage Problem (ASCP) as follows:

Definition 1. *Given a rectangular region R of area A and mobile sensors with limited memory that are capable of moving at a speed of v meters/second, find a plan of movement of mobile sensors that ensures a sweep period less than or equal to T and minimizes the number of mobile sensors required.*

As sensors have limited memory, we must incorporate this factor while optimizing the number of mobile devices. Later, we show that by increasing the memory capacity, the number of mobile devices used can be reduced.

8.4 Algorithm for Sweep Coverage

Our approach to solving ASCP is as follows. We find a set of points $P = \{p_1, p_2, \dots p_n\}$, where a sensor senses the data about itself. We must ensure that every point in the region R is within the coverage of at least one $p_k \in P$. Then we find a tour traversing the points in P and the BS. Finding an optimal tour is equivalent to solving the traveling salesman problem (TSP). Even though TSP is NP-complete, when the points in P follow a regular pattern, one can find the optimal tour easily. Let L be the length of the tour. The time required to complete the tour by a mobile device is $\tau = L/v$. If there are m mobile devices, each starting from BS at intervals of τ/m and following the same trajectory, then each point in P will be traversed at intervals τ/m. Since the sweep period is T, we must ensure $\tau/m \leq T$. So, the minimum number of mobile devices needed to ensure a sweep period of T is $\lceil \tau/T \rceil = \lceil \frac{L}{v \times T} \rceil$. We denote this minimum by $m_L = \lceil \frac{L}{v \times T} \rceil$. Assuming each mobile device completes traversing all the points in P before returning to BS, each mobile device has to carry the information about $|P|$ locations to the BS and thus needs a memory capacity $M|P|$ assuming each locations, data is of M bytes. From now onwards, $|P|$ will represent the memory requirements of the mobile device as M is constant. So, to minimize m, we need to minimize L, and at the same time, ensure that $|P|$ is not very high.

8.4.1 Hexagonal tessellation

To minimize $|P|$, we employ the following approach to find the set P. We place the points in a row with a horizontal spacing Δx, $\Delta x < 2r_{\mathrm{s}}$. Let the coordinates of two such consecutive points be $p_1 = (a, y)$ and $p_2 = (a + \Delta x, y)$. If we draw circles of radii r_{s} centered at p_1 and p_2, they would intersect at $(a + \Delta x/2, y \pm h)$ where $h = \sqrt{r_{\mathrm{s}}^2 - \Delta x^2/4}$. If the points in the next row are also placed with horizontal gap Δx, at positions $(a + \Delta x/2, y')$, $(a + 3\Delta x/2, y')$, and so on, to ensure no gap in coverage y' must not differ from y by more than $h + r_{\mathrm{s}}$. Thus spacing between the rows should be at most $h + r_{\mathrm{s}}$. If the width and height of the region are W and H, the number of points in a row is approximately $W/\Delta x$, and the total number of such rows is approximately $H/(h + r_{\mathrm{s}})$. Thus the total number of points $|P| \geq f(\Delta x) = \frac{WH}{\Delta x(r_{\mathrm{s}} + \sqrt{r_{\mathrm{s}}^2 - \Delta x^2/4})}$. To find the value of Δx that will minimize $f(\Delta x)$, we differentiate $f(\Delta x)$ with respect to Δx and equate it to 0. Then we get $\Delta x = \sqrt{3}r_{\mathrm{s}}$. So, for optimal placement, the horizontal spacing Δx should be $\sqrt{3}r_{\mathrm{s}}$. The corresponding h is $r_{\mathrm{s}}/2$. Hence the gap between rows is $h + r_{\mathrm{s}} = 3r_{\mathrm{s}}/2$.

Lemma 8.1. *The placement of points following the above rule will generate a set P such that $|P| > \frac{2A}{3\sqrt{3}r_s^2}$.*

Proof. The above placement scheme will give a tessellation of the region with a set of regular hexagons centered at the points of P. The side of the hexagon is of length r_{s}. The area of each hexagon is $3\sqrt{3}r_{\mathrm{s}}^2/2$, and they together have to cover the region of area A. Hence the total number of points in P must be at least $\frac{2A}{3\sqrt{3}r_{\mathrm{s}}^2}$.

In an actual case, it will be slightly more than $\frac{2A}{3\sqrt{3}r_{\mathrm{s}}^2}$ because, for the points lying near the boundary of R, some part of the hexagons would be outside R.

8.4.1.1 Hamiltonian tour traversing P

If the points in P occupy an even number of rows, constructing the tour is straightforward. We start from a point $p \in P$ at the bottom-left corner of R and move horizontally toward the right to the rightmost point. Then, we go up to the next row and traverse this row from right to left without visiting the

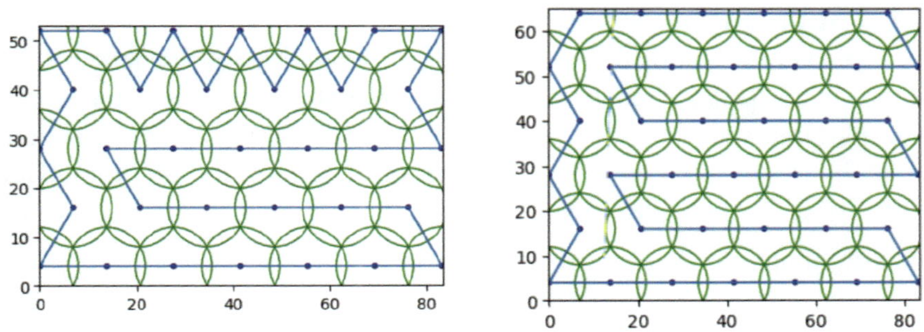

(a) The region is divided into odd number of rows

(b) The region is divided into even number of rows

Figure 8.1 Hamiltonian tour with hexagonal tessellation.

leftmost point. Thus, we move in a zigzag fashion toward the last row, leaving out the leftmost point in each row. When we reach the leftmost corner in the top row, we move downwards, traversing the leftmost points of each row until we reach the starting vertex. The minimum distance between any two points in P is $\sqrt{3}r_s$, and thus the optimal Hamiltonian tour must be of length at least $|P|\sqrt{3}r_s$. In the constructed path, the distance between consecutive points is always $\sqrt{3}r_s$, and we visit total $|P|$ points. Hence, this Hamiltonian tour is optimal for the given set P.

If the number of rows is odd, say $2k + 1$, we proceed as before and complete up to the rightmost point in row $2k - 1$. Then, we visit the points in the last two rows in a zigzag fashion until we reach the leftmost corner and then move downwards. Figure 8.1 shows the Hamiltonian tours for even and odd numbers of rows in P.

8.4.1.2 Approximation ratio for tour length

The work in [3] shows that the optimal tour length L_{opt} to cover a region of area A is at $\frac{A}{2r_s}$. If A is much larger than the area of each hexagon $3\sqrt{3}r_s^2/2$, we can approximate the number of points to be $\frac{2A}{3\sqrt{3}r_s^2}$, as the area of the hexagons at the boundary lying outside R would be much smaller compared to A. Then, the length of the Hamiltonian tour is $L = \frac{2A}{3\sqrt{3}r_s^2} \times \sqrt{3}r_s = \frac{2A}{3r_s}$. Thus, hexagonal tessellation will have an approximation ratio $L/L_{opt} = 4/3$.

We denote by M_o the optimal number of points in P that covers the region R of area A. M_o is expressed as

$$M_o = \frac{2A}{3\sqrt{3}r_s^2}. \tag{8.1}$$

8.4.1.3 Tour length for square grid

We can also find a tour by dividing the region into squares of side length $\sqrt{2}r_s$ and the set of points P consists of all the centers of each such square as a sensor placed at the center of a square will cover the entire square. Assuming W and H are divisible by $\sqrt{2}r_s$, the number of squares are $A/(2r_s^2)$. Thus cardinality of P is $A/(2r_s^2)$ which is more than M_o as $\frac{2}{3\sqrt{3}} < 1/2$. So, the length of an optimal TSP traversing all points of P is at least $|P| \times \sqrt{2}r_s = \frac{A}{\sqrt{2}r_s}$. Thus, a square grid division can never give us an approximation ratio smaller than $\sqrt{2}$. Thus, a square grid division for sweep coverage is inferior to a hexagonal division in tour length and memory requirements.

8.4.2 Memory tour length trade-off

Even though we have established the superiority of hexagonal division, we can reduce the tour length further by placing the points in a row at closer intervals. The tour has a total upward movement of approximately $2H/\sqrt{3}$ (if the lines incline at an angle of $60°$) or H (if the line goes straight up). The same goes for total downward movement. Hence, the only way to reduce the tour length is to decrease the number of horizontal lines. The number of rows will decrease if we increase the spacing $r_s + \sqrt{r_s^2 - \Delta x^2/4}$ (should be less than $2r_s$) between rows by making Δx smaller. However, it would increase the cardinality of the set P and thus the memory requirement of the mobile devices. If Δx is close to zero, the spacing between rows is almost $2r_s$, and we can reach near optimality in terms of tour length, but the cardinality of P would almost be infinite.

Since we cannot have indefinitely large memory, we use Algorithm 8.1 to determine the number of rows rc as follows. We use Δy to denote the spacing between rows. Leaving out a $r_s/2$ wide strip above the top row and below the bottom row, and assuming Δy to be $3r_s/2$, we find the row count as follows :

If $H - r_s$ is divisible by $3r_s/2$, then number of rows is $\frac{H-r_s}{3r_s/2} + 1$. If $H - r_s$ is not divisible by $3r_s/2$, the number of rows row that will keep the spacing between rows approximately $3r_s/2$ is

$$\text{row} = \left\lfloor \frac{H - r_s}{3r_s/2} \right\rfloor + 1. \tag{8.2}$$

We want to keep the number of rows even because as we try to increase Δy by decreasing Δx, and the number of rows is odd, the overhead due to moving back and forth between the last two rows will be too high (decreasing

Algorithm 8.1: Area sweep coverage.

Input : W: Width of the region; H: Height of the region; r_s: Sensing range of the sensor, v: Speed of mobile sensor nodes, T: Sweep period; α_{max}: Maximum allowable memory ratio
Output: $|P|$: Memory requirement; L: Length of a Hamiltonian tour; m: Number of mobile sensors

```
begin
    rc ← ⌊(H − r_s)/((3/2) r_s)⌋ + 1;
    if rc is odd then
        rc ← rc − 1;
    Δy ← (H + 2r_s)/(rc + 1);
    while Δy < 2r_s do
        compute α by eqn (8.4), (8.5);
        if α > α_max then
            break;
        rc ← rc − 2;
        Δy ← (H + 2r_s)/(rc + 1);
    rc ← rc + 2;
    P ← Placement(W, H, r_s, rc);
    Calculate the Length L of Hamitonian tour traversing each p ∈ P;
    Calculate the total mobile sensor requirement m = ⌈L/(T × v)⌉;
    return (|P|, L, m)
```

Δx means there would be many more points in a row). Keeping the number of rows even will also make the tour simple. So, in our algorithm, the row count rc is given by

$$rc = \begin{cases} row, & \text{if } row \text{ is even} \\ row - 1, & \text{otherwise} \end{cases}.$$

where row is obtained by eqn (8.2).

For a given rc, the value of Δy is given by the following equation:

$$\Delta y = \frac{H + 2r_s}{rc + 1}. \tag{8.3}$$

Eqn (8.3) is derived as follows. For a given Δy, we can leave a space of $\Delta y - r_s$ above the topmost row and $\Delta y - r_s$ below the bottom most row without creating any coverage gap. So $H - 2(\Delta y - r_s) = (rc - 1)\Delta y \implies \Delta y = \frac{H+2r_s}{rc+1}$.

For a given Δy, the horizontal spacing Δx can be found by the equation

$$\Delta x = 2\sqrt{2\Delta y \times r_s - \Delta y^2}. \tag{8.4}$$

We derive eqn (8.4) as follows:

$$\begin{aligned}
\Delta y &= r_s + \sqrt{r_s^2 - \Delta x^2/4} \\
\Longrightarrow (\Delta y - r_s)^2 &= (r_s^2 - \Delta x^2/4) \\
\Longrightarrow \Delta x^2/4 &= 2\Delta y \times r_s - \Delta y^2 \\
\Longrightarrow \Delta x &= 2\sqrt{2\Delta y \times r_s - \Delta y^2}.
\end{aligned}$$

The number of points per row is approximately $W/\Delta x$. Hence, for a given Δy, the cardinality of the set P is approximately $rc \times W/\Delta x$. So, the memory is more than the optimal M_o by a factor α given by

$$\alpha = \frac{rc \times W}{\Delta x \times M_o}. \tag{8.5}$$

The value of α depends only on row count rc as Δx is a function of Δy that is again a function of rc. The other terms – W and M_o are constants for a given region and sensing range.

We fix an upper bound on α say α_{max} and keep on decreasing rc by 2, as long as Δy remains less than $2r_s$ and α does not exceed α_{max}.

8.4.3 Placement of points for sweep coverage

Once we arrive at a value of the number of rows rc, we determine Δy and Δx by applying eqn (8.3) and (8.4). Then Algorithm 8.2 places the points row by row as follows. For the first row, the points are placed at $(0, y)$, $(\Delta x, y)$, $(2\Delta x, y)$ and so on where y is $\Delta y - r_s$. For the next row, y increases by Δy, and x starts from $\Delta x/2$. That way, there will be no gap in coverage at the left end. To find the maximum value of x at each row, we note that two circles of radius r_s centered at (x, y) and $(x, y + \Delta y)$ would intersect at $x \pm \sqrt{r_s^2 - \Delta y^2/4}$. To avoid a coverage gap, we want $x + \sqrt{r_s^2 - \Delta y^2/4}$ to be equal to W. Thus the maximum x in a row is $xmax = W - \sqrt{r_s^2 - \Delta y^2/4}$. So, in a row, we keep increasing x as long as x is below $W - xmax$. Then we put the rightmost point at x equal to $xmax$. To avoid putting points too close, we check that $xmax$ is more than the immediately preceding point's x coordinate by a small constant ϵ.

Figure 8.2 illustrates a tour obtained when $W = 60$ m, $H = 30$ m, and $r_s = 4$ m. The blue circles are the points in P where mobile sensors sense data within a circle of radius r_s. The tour length is 270 m, whereas the optimal tour length is at least $60 \times 30/8 = 225$ m. The approximation ratio for tour length is 1.20. The memory required is 72 ($1.67 M_o$). For a sweep period of 30 seconds and a speed of mobile device 1 m/s, the number of mobile devices needed is 9. In Figure 8.2, the mobile devices at a specific point in time are shown as red circles.

Algorithm 8.2: Placement.

```
Input  : W: Width, H: Height, r_s : Sensing radius, rc: Row count;
Output: P: Set of points
begin
    P ← ∅;
    Δy ← (H+2r_s)/(rc+1);
    y ← Δy − r_s;
    Compute Δx by Equation 8.4;
    xmax ← W − sqrt(r_s^2 − Δy^2/4);
    for (row = 0; row < rc; row++) do
        x ← 0.5Δx × (row mod 2);
        while x < xmax do
            add (x, y) to P; x ← x + Δx;
        if xmax − (x − Δx) > ε then
            add (xmax, y) to P;
        y ← y + Δy;
    return P
```

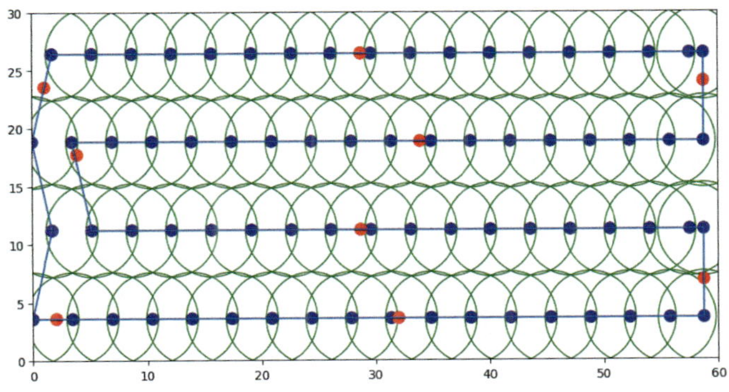

Figure 8.2 Tour formed with proposed algorithm along with placement of mobile sensors marked in red.

8.4.3.1 Complexity analysis

In the while loop of Algorithm 8.1, rc starts from $\frac{2(H-r_s)}{3r_s}+1$ and decreases in steps of 2. It cannot decrease beyond $\frac{H-r_s}{2r_s}+1$. So the number of iterations in the while loop is $O(\frac{H-r_s}{r_s}\frac{2/3-1/2}{2}) = O(\frac{H-r_s}{12r_s})$. For example, with $H =$ 200 m and $r_s = 7$ m, $\frac{H-r_s}{12r_s} = 2.3$, and there will be two iterations where rc goes from 18 to 14. Algorithm 8.2 is executed only once and its complexity is $O(|P|)$, and $|P|$ is at most $\alpha_{max}M_o = \alpha_{max}\frac{2A}{3\sqrt{3}r_s^2}$.

8.5 Analysis of Results

In our experiment, we have taken the surveillance region of different dimensions some square and some rectangular. We computed the tour length assuming two different mobile sensing radii, 7 and 9 meters. The sweep period and velocity of the mobile sensor nodes are 30 seconds and 5 m/s, respectively. Table 8.1 shows the summarized results in terms of tour length and memory requirements for square subdivisions, and Table 8.2 shows the summarized results in terms of tour length and memory requirements using our algorithm (Algorithm 8.1). We got the tour length approximation ratio by dividing the tour length by $T_o = \frac{A}{2r_s}$. Dividing $|P|$ by $M_o = \frac{2A}{3\sqrt{3}r_s^2}$ gives the memory approximation ratio. The tables show that the proposed algorithm has a much lower approximation ratio for tour length than the method based on a square grid [3]. Also, the proposed method works for any dimension of the rectangular area without requiring the dimensions to be integral multiples of $\sqrt{2}r_s$. We have shown that with the horizontal spacing $\sqrt{3}r_s$ between the points in P, we get a hexagonal tessellation that gives a tour of length 4/3 times the optimal. However, in the proposed method, we have decreased

Table 8.1 Tour of mobile sensor nodes for square subdivision.

Sensing radius (meters)	Dimension (meters)	Results obtained with square sub-division of region [3]				
		Tour length (meters)	Tour length approx ratio	Memory reqd.	Memory reqd. approx ratio	# of mobile sensors reqd.
7	200 × 200	4365.68	1.53	441	1.40	30
	300 × 150	4910.15	1.53	496	1.41	33
	350 × 200	7488.12	1.49	756	1.38	50
9	200 × 200	3263.62	1.47	256	1.35	22
	300 × 150	3670.91	1.47	288	1.35	25
	350 × 200	5707.38	1.47	448	1.35	39

Table 8.2 Tour of mobile sensor nodes for the proposed method.

Sensing radius (meters)	Dimension (meters)	Result after first iteration		Final result		
		Memory approx ratio	Tour length approx ratio	Memory approx ratio	Tour length approx ratio	# of mobile sensors reqd.
7	200 × 200	1.09	1.29	1.25	1.19	23
	300 × 150	1.05	1.33	1.24	1.16	25
	350 × 200	1.06	1.28	1.24	1.16	39
9	200 × 200	1.11	1.31	1.45	1.18	18
	300 × 150	1.13	1.24	1.13	1.24	21
	350 × 200	1.08	1.29	1.43	1.14	30

the horizontal spacing from $\sqrt{3}r_s$, reduced the number of rows, and got an approximation ratio for tour lengths in the range of 1.14 to 1.24.

Table 8.2 shows the memory approximation ratio and tour length approximation ratio for the first iteration (when horizontal spacing is approximately $\sqrt{3}r_s$) and the corresponding values for the final result. In most cases, the final result has a higher approximation ratio for memory and a lower approximation ratio for tour lengths. Only for one case (dimension 300 × 150 and $r_s = 9$), there is no change between the first iteration and the final result. The reason is that the spacing between the rows is already much more than $3/2r_s$ in the first iteration with $rc = 10$. If we reduce rc to 8, $\Delta y = (H + 2r_s)/(rc + 1)$ exceeds $2r_s$, and hence we cannot reduce rc. In this case, the memory approximation ratio is small, and the tour length approximation ratio is large. For other cases, we can get Δy close to $2r_s$, thereby making the points in a row closer together (increasing memory requirements) and the number of rows rc smaller (reducing the tour length and tour length approximation ratio).

8.6 Conclusion

While designing the plan for sweep coverage, one can opt for hexagonal tessellation with optimal memory but tour length with an approximation ratio of 4/3. On the other hand, one can opt for shorter tours (approximation ratio close to one and a smaller number of mobile sensors) at the expense of more memory requirements at the mobile nodes. For future work, we can consider keeping the memory requirements down while optimizing the tour length by dividing the surveillance region into nonoverlapping zones and designing a tour for each zone. Since a mobile device will need to carry the information about a fraction of the set of points in P, it will reduce memory requirements. A centrally placed BS makes such division into zones simple, but a randomly placed BS will create irregularly shaped zones. In such a scenario, it will be interesting to see how it affects the number of mobile devices.

References

[1] Mohammed Boudali, Mustapha Reda Senouci, Mohamed Aissani, and Walid-Khaled Hidouci. Activities scheduling algorithms based on probabilistic coverage models for wireless sensor networks. *Annals of Telecommunications*, 72(3):221–232, 2017.

[2] Barun Gorain and Partha Sarathi Mandal. Point and area sweep coverage in wireless sensor networks. In *2013 11th International Symposium and Workshops on Modeling and Optimization in Mobile, Ad Hoc and Wireless Networks (WiOpt)*, pages 140–145. IEEE, 2013.

[3] Barun Gorain and Partha Sarathi Mandal. Approximation algorithms for sweep coverage in wireless sensor networks. *Journal of parallel and Distributed Computing*, 74(8):2699–2707, 2014.

[4] Andrew Howard, Maja J Matarić, and Gaurav S Sukhatme. Mobile sensor network deployment using potential fields: A distributed, scalable solution to the area coverage problem. In *Distributed autonomous robotic systems 5*, pages 299–308. Springer, 2002.

[5] Mo Li, Weifang Cheng, Kebin Liu, Yuan He, Xiangyang Li, and Xiangke Liao. Sweep coverage with mobile sensors. *IEEE Transactions on Mobile Computing*, 10(11):1534–1545, 2011.

[6] Emanuele Magrini, Fabrizio Flacco, and Alessandro De Luca. Estimation of contact forces using a virtual force sensor. In *2014 IEEE/RSJ International Conference on Intelligent Robots and Systems*, pages 2126–2133. IEEE, 2014.

[7] Madana Srinivas, Praveen Kumar Donta, and Tarachand Amgoth. Efficient algorithms for point and area sweep–coverage in wireless sensor networks. In *2021 Sixth International Conference on Wireless Communications, Signal Processing and Networking (WiSPNET)*, pages 315–320. IEEE, 2021.

[8] Massimo Vecchio and Roberto López-Valcarce. Improving area coverage of wireless sensor networks via controllable mobile nodes: A greedy approach. *Journal of network and computer applications*, 48:1–13, 2015.

[9] Yang Wang, Shuang Wu, Zhiyin Chen, Xiaofeng Gao, and Guihai Chen. Coverage problem with uncertain properties in wireless sensor networks: A survey. *Computer Networks*, 123:200–232, 2017.

[10] Qianqian Yang, Shibo He, Junkun Li, Jiming Chen, and Youxian Sun. Energy-efficient probabilistic area coverage in wireless sensor networks. *IEEE Transactions on Vehicular Technology*, 64(1):367–377, 2014.

9

Hardware Implementation of a Direction of Arrival Estimation Algorithm using LU Decomposition

Nizar Tayem[1], AbuMuhammad Moinuddeen[1], Sebastian Mayfield[1], Ahmed A. Hussain[2], and Redha M. Radaydeh[1]

[1]Department of Engineering & Technology, Texas A & M University, United States
[2]Department of Electrical Engineering, Prince Mohammad University, Saudi Arabia
E-mail: nizar.tayem@tamuc.edu; amoinuddeen@leomail.tamuc.edu; smayfield3@leomail.tamuc.edu; ahussain1@pmu.edu.sa; redha.radaydeh@tamuc.edu

Abstract

This chapter introduces an innovative approach for estimating the direction of arrival (DOA) of multiple RF noncoherent sources in a uniform linear array (ULA). The proposed method utilizes a Capon-like estimation algorithm and incorporates LU decomposition to enhance the accuracy of DOA estimation while significantly reducing computational complexity compared to existing methods like the Capon method. Notably, the proposed method does not require prior knowledge of the number of sources. To validate its effectiveness, the proposed method undergoes validation through both software simulations and practical experimentation on a prototype testbed constructed using a software-defined radio (SDR) platform and GNU radio software. The results obtained from MATLAB simulations and real-time experiments provide compelling evidence of the proposed method's efficacy.

Keywords: DOA estimation, real-time validation, software defined radio, computational complexity, Capon method, GNU radio

DOI: 10.1201/9788743801108-10

9.1 Introduction

Direction of arrival (DOA) estimation has become increasingly important due to its use in applications in fields such as 5G wireless communication, radar, radio astronomy, sonar, navigation, aiding search and rescue missions of people in distress or imminent danger, detecting targets and target communications in hostile/wartime regions, enable beamforming toward intended receivers and much more [1]. Over the years, due to its many applications and the need for improvement in accuracy and efficiency, there has been a considerable increase in research being conducted [2]. Of the many DOA estimation algorithms reported in the literature, the two most widely researched and implemented algorithms for DOA are the high-resolution multiple signal classification (MUSIC) [3] and estimation of signal parameters via rotational invariant techniques (ESPRIT) [4]. MUSIC is a subspace-based method that utilizes the eigenstructure of the covariance matrix of the received signal to estimate the directions of arrival of multiple signals while ESPRIT is another subspace-based method, but it relies on the rotational invariance property of the signal subspace. It is often considered more computationally efficient than MUSIC because it does not require a spectral search.

Hardware implementation and real-time experimental validation of DOA estimation algorithms are of critical importance before these algorithms can be deployed in real-world practical applications. While several works have reported hardware implementations and experimental validation, these have been developed mostly using proprietary software. Most real-time research implementations using hardware utilize the LabView software developed by National Instruments (NI) along with NI's software-defined radios as there is readily available documentation and extensive and quick support from NI. However, NI equipment and software are not free and open-source. A rising alternative is GNU Radio Companion (GRC) [5], which is a "free and open-source software development toolkit that provides signal processing blocks to implement software radios" [6]. Initially, GRC was mostly used by radio hobbyists but now, GRC is also used by researchers as it can be used to program SDRs, as well as develop simulations, a variety of in-built blocks, and Python-embedded blocks which can be programmed as needed. However, GRC is still not widely used in research, and thus, available documentation and literature are relatively quite limited.

Popular algorithms for DOA, like the commonly used MUSIC and ESPRIT and their variant implementations [7–18] mostly use eigenvalue decomposition (EVD) or singular value decomposition (SVD) to separate the signal space from the noise space in order to calculate the estimated DOA. The decomposition methods are computationally intensive and are difficult to apply in hardware implementations (real-time). QR decomposition is much better for real-time implementations in terms of speed, computational intensity, and accuracy [19]. However, QR decomposition is still considerably computationally intensive and very slow when implemented in real-time using hardware. The work reported in [20] validates a channelized version of the MUltiple SIgnal Classification (MUSIC) algorithm for DoA estimation using experimentally acquired high chirp-rate LFM signal data. This technique avoids pre-processing and combines spatial pseudospectra from individual frequency channels incoherently to obtain a single accurate estimate. Another work in [21] presents an FPGA implementation and real-time experimental verification of a proposed direction of arrival (DoA) estimation algorithm using LDL decomposition. The algorithm is implemented on a Xilinx FPGA using LabVIEW software, and its real-time verification is conducted using the National Instruments (NI) PXI platform. In [22], the authors employed blind separation techniques to estimate the direction of arrival (DOA) using a smart linear antenna array system, implemented the proposed algorithm on an FPGA, and validated it experimentally. However, all these methods have the drawback of higher computational complexity as they require either the eigenvalue decomposition (EVD) of the covariance matrix or the singular value decomposition (SVD) of the received data matrix.

In contrast, using LU decomposition to extract the signal and noise subspaces significantly reduces the computational complexity of the DOA algorithm [23]. LU decomposition factorizes a matrix A such that $A = LU$ where L is the lower triangular matrix and U is the upper triangular matrix [23]. LU decomposition has good accuracy with much better cost performance compared to QR decomposition [20]. LU decomposition's relatively low computational intensity when computing the inverse of triangular matrices, speed, and good accuracy make it better suited for real-time hardware implementations [24].

This chapter presents the findings of the implementation of our Capon-like method for one-dimensional (1D) DOA estimation which utilizes LU decomposition. This chapter presents the hardware implementation and real-time experimentation validation of the proposed DOA estimation algorithm which is based on a Capon-like method utilizing LU decomposition. The experimental validation is performed on a testbed built using the National Instruments (NI) SDR platform [28] and a four-element ULA for the estimation of DOA in one dimension. The key advantages of the proposed method are: (1) it has much lower computational complexity compared to existing methods such as Capon [25–26], (2) it does not require computationally complex EVD or SVD as it uses LU decomposition, (3) the number of sources is not required to be known by the algorithm in advance, and (4) its low computational complexity makes it suitable for deployment in practical real-time applications.

The main sections in the rest of the chapter are organized as: Section 9.2 will explain the system model and the proposed algorithm, Section 9.3 will present the complexity analysis of the proposed algorithm, Section 9.4 will explore the results of MATLAB simulations based on the proposed algorithm presented along with the results of the real-time implementation using GNU Radio Companion, and finally, the conclusions from the research work are drawn in Section 9.5.

9.2 System Model for DOA Estimation

Consider a scenario involving k narrowband uncorrelated sources situated in the far-field region of a uniform linear array (ULA). The ULA is composed of M antennas positioned along the x-axis. The separation between adjacent elements is $\lambda/2$, where λ represents the wavelength of the incident source. For instance, Figure 9.1 depicts a ULA with four antenna elements, with two sources located at θ_1 and θ_2, respectively, impinging on it.

The observation signal received from the antenna elements placed on the x-axis in the case of M antennas and k sources can be expressed as

$$\boldsymbol{x}(t) = \boldsymbol{A}_x(\theta)\boldsymbol{S}(t) + \boldsymbol{N}_x(t) \tag{9.1}$$

where $\boldsymbol{A}_x(\theta)$ is the manifold matrix of the x-axis.

$$\boldsymbol{A}_x(\theta) = \left[\boldsymbol{a}_x(\theta_1), \ldots, \boldsymbol{a}_x(\theta_k)\right], \tag{9.2}$$

$$\begin{aligned}\boldsymbol{a}_x(\theta_k) &= \left[a_{x,1}(\theta_k), a_{x,2}(\theta_k), \ldots, a_{x,M}(\theta_k)\right]^T \\ &= \left[1, e^{\frac{j2\pi d\cos\theta_k}{\lambda}}, \ldots, e^{\frac{j2\pi Md\cos\theta_k}{\lambda}}\right]^T.\end{aligned} \tag{9.3}$$

$\boldsymbol{S}(t) = [s_1(t), s_2(t), \ldots, s_k(t)]^T$ is a $K \times 1$ signal vector received from K sources.

$\boldsymbol{N}_x = [n_{x,1}(t), n_{x,2}(t), \ldots, n_{x,M}(t)]^T$ and with dimension $M \times 1$ for a single snapshot are the complex vectors of the noise received by the x-axis for L snapshots,

$\boldsymbol{S}(t) \in C^{K\times L}$, $\boldsymbol{N}_x(t) \in C^{M\times L}$ and $\boldsymbol{x}(t) \in C^{M\times L}$.

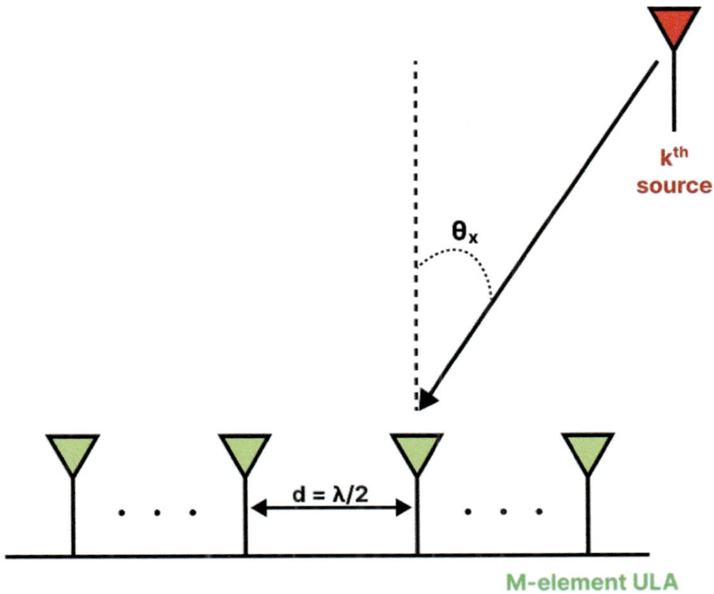

Figure 9.1 System model employing a uniform linear array.

9.3 Proposed Algorithm

In this particular section, we present a novel algorithm for estimating the direction of arrival (DOA). To initiate the algorithm, we create the correlation matrix using the received signal vector $\boldsymbol{x}$(t). Mathematically, the correlation matrix can be computed as

$$\boldsymbol{R}_{xx} = E\left(\boldsymbol{x}(t)\boldsymbol{x}(t)^H\right) = \boldsymbol{A}(\theta)\boldsymbol{R}_s\boldsymbol{A}^H(\theta) + \boldsymbol{R}_N. \tag{9.4}$$

As the sources are uncorrelated, the source signal covariance matrix $\boldsymbol{R}_s$ becomes a nonsingular matrix with a rank equivalent to the number of sources. Additionally, $\boldsymbol{R}_N = \sigma^2\boldsymbol{I}$ is the noise covariance matrix, where $\boldsymbol{I}\epsilon\boldsymbol{C}^{(M\times M)}$ is the identity matrix. In practice, the covariance matrix can be computed using L samples as follows:

$$\widehat{\boldsymbol{R}}_{xx} = \frac{1}{L}\sum_{l=1}^{L}\boldsymbol{x}\left(t_l\right)\boldsymbol{x}^H\left(t_l\right). \tag{9.5}$$

The covariance matrix $\widehat{\boldsymbol{R}}_{xx} \in C^{(M\times M)}$ is a nonsingular square matrix and it can be factored using LU factorization in which all diagonal elements of the upper triangular matrix are set to one. Calculating the LU matrices requires $O\left(M^3/3\right)$ operations using Crout's method. LU decomposition is a product of a matrix $\boldsymbol{L} \in \boldsymbol{C}^{(M\times M)}$ and a matrix $\boldsymbol{U} \in \boldsymbol{C}^{(M\times M)}$, such that

$$\widehat{\boldsymbol{R}}_{xx} = \boldsymbol{L_x U_x}. \tag{9.6}$$

The estimated covariance matrix $\widehat{\boldsymbol{R}}_{\mathrm{Xx}}$ can be expressed as

$$\widehat{\boldsymbol{R}}_{xx} = \begin{bmatrix} r_{11} & r_{12} & r_{13} & r_{14} & \cdots & r_{1M} \\ r_{21} & r_{22} & r_{23} & r_{23} & \cdots & r_{2M} \\ r_{31} & r_{32} & r_{33} & r_{34} & \cdots & r_{3M} \\ r_{41} & r_{11} & r_{11} & r_{44} & \cdots & r_{4M} \\ \vdots & \vdots & \vdots & \vdots & \ddots & \vdots \\ r_{M1} & r_{M2} & r_{M3} & r_{M4} & \cdots & r_{MM} \end{bmatrix}. \tag{9.7}$$

$\boldsymbol{L}_x$ is the lower triangular matrix which can be expressed as

$$\boldsymbol{L_x} = \begin{bmatrix} l_{11} & 0 & 0 & 0 & 0 & 0 \\ l_{21} & l_{22} & 0 & 0 & 0 & 0 \\ l_{31} & l_{32} & l_{33} & 0 & 0 & 0 \\ l_{41} & l_{42} & l_{43} & l_{44} & 0 & 0 \\ \vdots & \vdots & \vdots & \vdots & \ddots & 0 \\ l_{M1} & l_{M2} & l_{M3} & \cdots & \cdots & l_{MM} \end{bmatrix}, \tag{9.8}$$

and the upper triangular matrix $\boldsymbol{U_x}$ can be expressed as

$$\boldsymbol{U_x} = \begin{bmatrix} 1 & u_{12} & u_{13} & u_{14} & \cdots & u_{1M} \\ 0 & 1 & u_{23} & u_{24} & \cdots & u_{2M} \\ 0 & 0 & 1 & u_{34} & \cdots & u_{3M} \\ 0 & 0 & 0 & 1 & \cdots & u_{4M} \\ 0 & 0 & 0 & 0 & \ddots & \vdots \\ 0 & 0 & 0 & 0 & \cdots & 1 \end{bmatrix}. \tag{9.9}$$

The covariance matrix $\widehat{\boldsymbol{R}}_{xx}$ with entries $r_{ij}, i = 1, \cdots, n, j = 1, \cdots, n$, will be used to calculate the entries $\boldsymbol{L}_x$ and $\boldsymbol{U_x}$ using Crout's method as follows:

$$l_{i1} = r_{i1}, \text{ for } i = 1, 2, \ldots, M \tag{9.10}$$

$$u_{1j} = \frac{r_{1j}}{l_{11}}, \text{ for } j = 1, 2, \cdots, M \tag{9.11}$$

for $j = 2, 3, \cdots, M$

$$l_{ij} = r_{ij} - \sum_{k=1}^{j-1} l_{ik} u_{kj} \text{ for } i = j, j+1, \cdots, M \tag{9.12}$$

$$u_{jk} = \frac{r_{jk} - \sum_{i=1}^{j-1} l_{ji} u_{ik}}{l_{ij}}, \text{ for } k = j+1, j+2, \cdots, M, \tag{9.13}$$

$$l_{nn} = r_{MM} - \sum_{k=1}^{n-1} l_{nk} u_{kn}. \tag{9.14}$$

To estimate the azimuth angle of multiple sources for DOA, the upper triangular matrix $\boldsymbol{U_x}$ is derived through LU decomposition of the covariance matrix $\widehat{\boldsymbol{R}}_{xx}$. This matrix is utilized in conjunction with a one-dimensional special spectrum function. By analyzing the peaks in the spectrum plot generated from the 1D special spectrum, the localization of multiple RF sources $[(\theta_1), \ldots, (\theta_k)]$ can be determined as

$$P_p(\theta) = \frac{1}{\boldsymbol{A}^H(\theta)\left(U_{\mathrm{x}}^{-1}\right)\left(U_{\mathrm{x}}^{-1}\right)^H \boldsymbol{A}(\theta)}. \tag{9.15}$$

This is different from the Capon method [25-26] which uses a covariance matrix when computing the spatial spectrum. This process can be formulated by this equation

$$P_c(\theta) = \frac{1}{\boldsymbol{A}^H(\theta)\left(\widehat{\boldsymbol{R}}_{xx}{}^{-1}\right)\boldsymbol{A}(\theta)}. \tag{9.16}$$

The number of incident sources directly corresponds to the number of peaks in the spectrum plot.

9.4 Complexity Analysis

Complex-valued operations were used as the basis for comparing the proposed method with existing methods. The comparison focused on major signal processing operations in both the proposed and Capon methods. In the proposed method, LU decomposition was employed to calculate the lower triangular matrix, L_x, and the upper triangular matrix, U_x, from the covariance matrix. The U_x matrix was then utilized to compute the spatial spectrum according to eqn (9.15). On the other hand, the Capon method directly utilized the covariance matrix as shown in eqn (9.16). This significant difference in operations led to reduced complexity in the proposed method.

For a system with M antenna elements, the proposed method required approximately $O\left(M^{\beta}/3\right)$ operations to calculate the LU decomposition and $O\left(M^{2}\right)$ operations to compute the inverse of the upper triangular matrix, U_{x}. In contrast, the Capon method required around $O\left(M^{\beta}\right)$ operations to calculate the inverse of the covariance matrix, $\widehat{\boldsymbol{R}}_{xx}$. The special structure of the U_{x} matrix, being an upper triangular matrix with all diagonal elements set to ones, contributed to the complexity reduction. Additionally, the inverse of the diagonal elements of U_{x} remained the same.

To illustrate, consider a scenario with a massive-MIMO system comprising 128 antenna elements. In this case, the proposed method required 715,434 operations, while the Capon method required approximately 193% more operations, totaling 2,097,152. This significant reduction makes the proposed method a better candidate for high-speed communication applications such as 5G and 6G technology. Therefore, it can be concluded that the proposed method exhibits lower computational complexity while delivering superior performance, as elaborated in the subsequent section.

9.5 MATLAB Simulation and Experimental Results

9.5.1 MATLAB simulation

The proposed method's accuracy in estimating the direction of arrival (DOA) has been verified through both MATLAB simulations and real-time experiments. The validation was conducted under various scenarios, and its performance was compared to the Capon method in the MATLAB simulation case. The simulation results are depicted in Figures 9.2 and 9.3 as spatial spectrum plots.

Figure 9.2 illustrates the spatial spectrum of the 1D DOA estimation for two noncoherent sources, comparing the proposed method with the Capon method. The simulation parameters included a total of 8 elements ($M = 8$), a signal-to-noise ratio of 5 dB (SNR $=$ 5 dB), 500 snapshots, and the two source signals located at $30°$ and $45°$. These plots clearly demonstrate that the proposed method accurately estimates the DOA by displaying distinct, sharp, and clear peaks corresponding to the source locations. In contrast, the Capon method exhibits spectral leakage and can only detect the source locations with less precision.

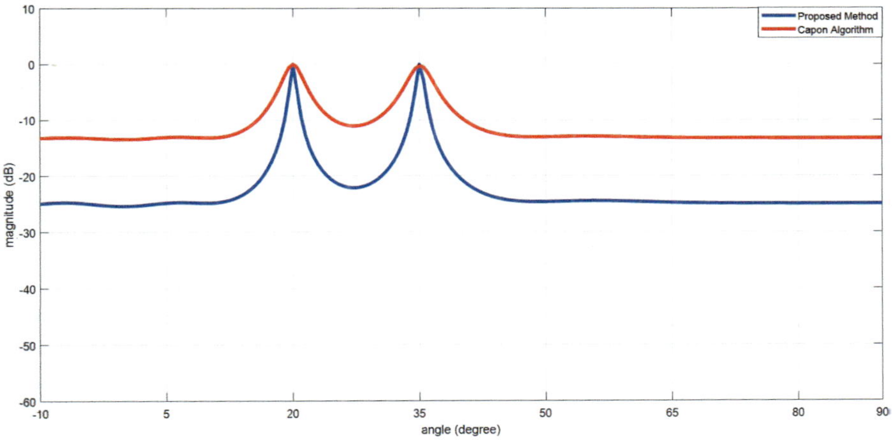

Figure 9.2 Spatial spectrum plot for two sources placed at (30°) and (45°) for the proposed method and Capon algorithm.

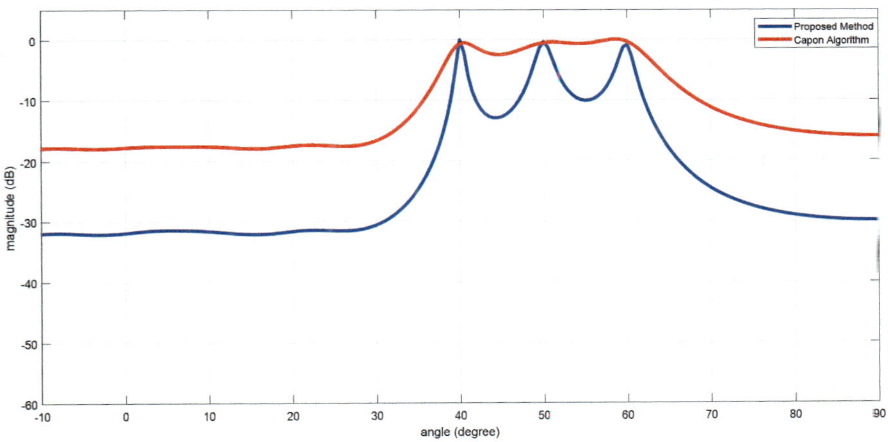

Figure 9.3 Spatial spectrum plot for three sources placed at (40°), (50°), and (60°) for the proposed method and Capon algorithm.

Figure 9.3 presents the spatial spectrum of the 1D DOA estimation for three noncoherent sources, again comparing the proposed method with the Capon method. The simulation parameters for this case were the same as before, with 8 elements ($M = 8$), a signal-to-noise ratio of 10 dB (SNR = 10 dB), 500 snapshots, and the three source signals located at 40°, 50°, and

60°. The plots clearly indicate that the proposed method accurately estimates the DOA, even when the source spacing is around 10°. On the other hand, the Capon algorithm fails to detect the locations of the sources accurately and can only identify one source at 40°.

It is evident from Figures 9.2 and 9.3 that the proposed method shows better resolution. This is due to the fact that LU decomposition in the proposed method reduces the effect of noise in the received data while the Capon method applies raw data without any signal processing steps.

9.5.2 Experimental setup

The setup for our real-time implementation of the estimation of the direction of arrival of 1 and 2 sources includes a CDA-2990 Octoclock [27], a waveform generator, two computers running Ubuntu 20.04 with GNU Radio Companion (GRC), UHD (universal driver), Python 3.10, and Vscode installed. The UHD (USRP hardware driver) establishes an interface between the USRPs and the computers which are then used to receive, transmit, and further program the USRPs using GNU Radio or GNU Radio Companion. Vscode facilitates programming in Python. One computer is solely used for transmitting signals and the other computer is used to receive the signals, synchronize the time and phases, and estimate the direction of arrival. The source transmits a signal of x_freq at x_rate using a NI-2901 USRP software-defined radio device [28] with one vert antenna connected to a vert antenna via a meter-long coaxial cable. The transmitted signal is received by a uniform linear array (ULA) which consists of four vert antennas connected to two NI-2901 USRPs (2 vert antennas per USRP) via meter-long coaxial cables. Both the transmitting and receiving arrays have been 3D-printed in-house. A CDA-2990 Octoclock and the waveform generator connected to it are used to synchronize the time on the receiving USRPs and to provide the pulse per second (PPS) and reference. Another USRP is used to relay a reference signal of 10 MHz for the phase correction process. The reference signal from the USRP is relayed to the receiving USRPs via meter-long coaxial cables and a PD1040 RF splitter. Diagrams of all the actual connections as seen in the lab and the schematic version of the setup can be found in Figures 9.4 through 9.8.

Figure 9.4 depicts the schematic diagram of the testbed setup for the DOA estimation of a single source.

Figures 9.5 and 9.7 show the actual testbed built in the lab for the DOA estimation of a single source using NI USRP-2901 [29]. Figure 9.6 shows the

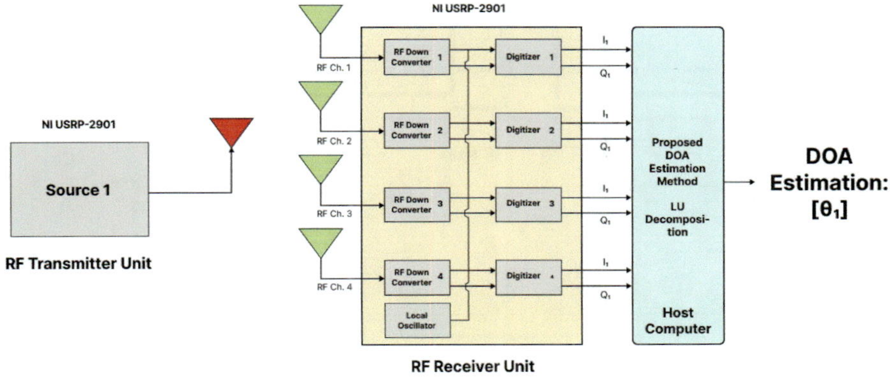

Figure 9.4 Schematic diagram of one source DOA estimation setup.

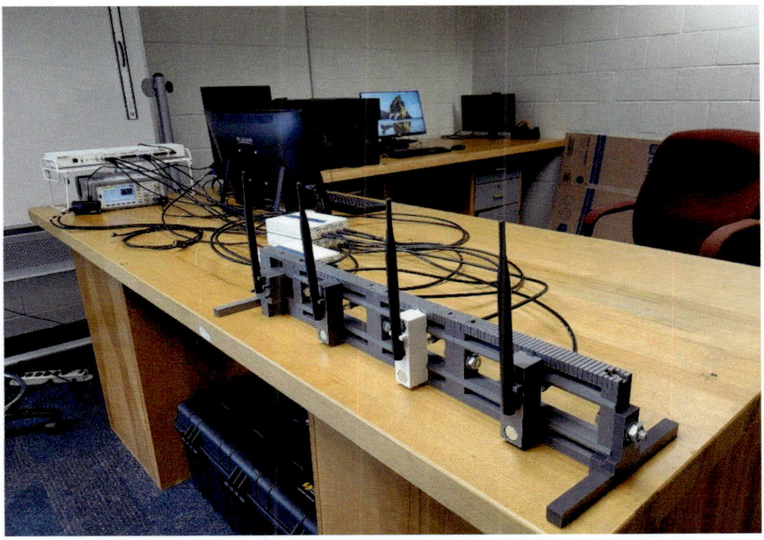

Figure 9.5 Four-element receiving antenna array.

connection between the USRPs and the host PC running the DOA estimation algorithm.

Figure 9.8 below shows the GNU Radio flowgraph for the implementation of the DOA estimation algorithm running on the host PC.

This real-time implementation of direction of arrival estimation in 1 dimension can be divided into the following steps: source signal generation, phase and time synchronization, and DOA estimation for 1 and 2 sources.

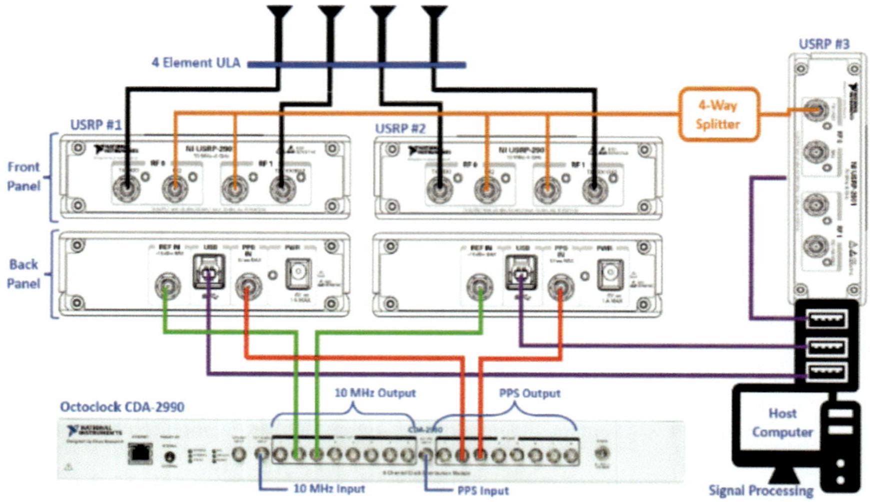

Figure 9.6 Connection diagram of USRP SDR receiver four-element ULA for real-time experiments.

Figure 9.7 Tesbed for DOA Estimation of a Single Source.

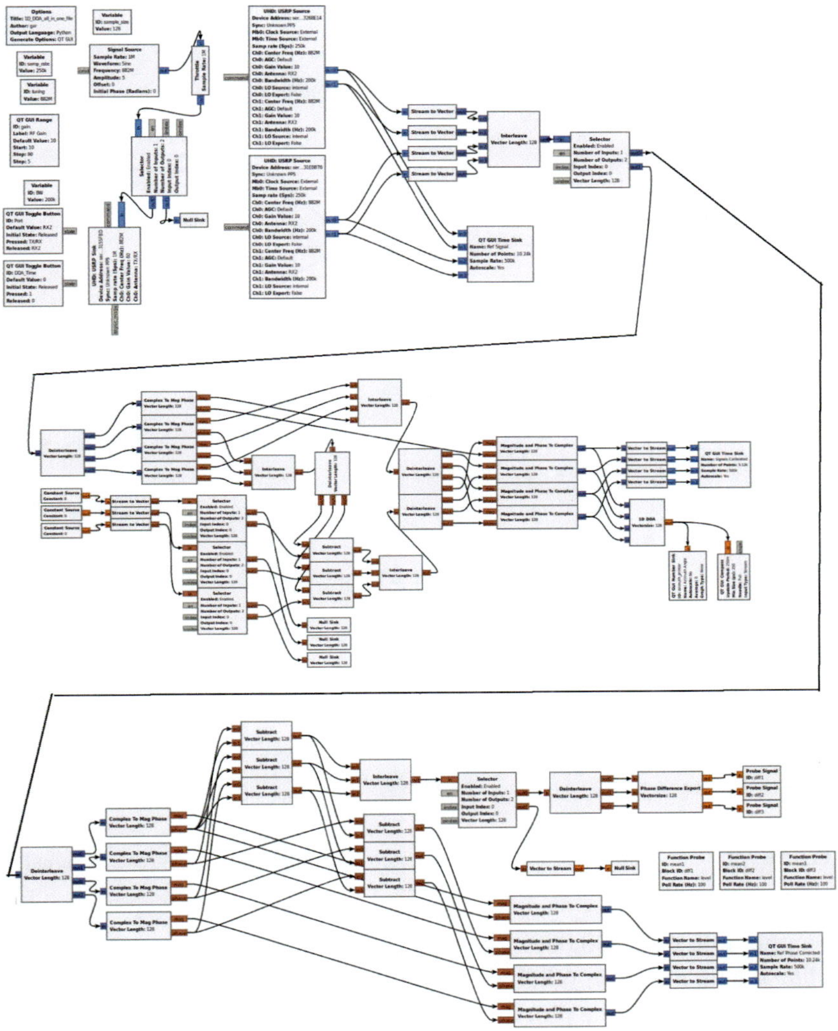

Figure 9.8 GNU Radio flowgraph.

9.5.3 Scenario 1: DOA estimation of one source

Step 1 – source signal generation:

To estimate the direction of arrival, we need a source. For our source, we use the *signal source block* along with the *UHD USRP sink block* to generate the source signal. An 882 MHz signal is generated by the *signal source block* at

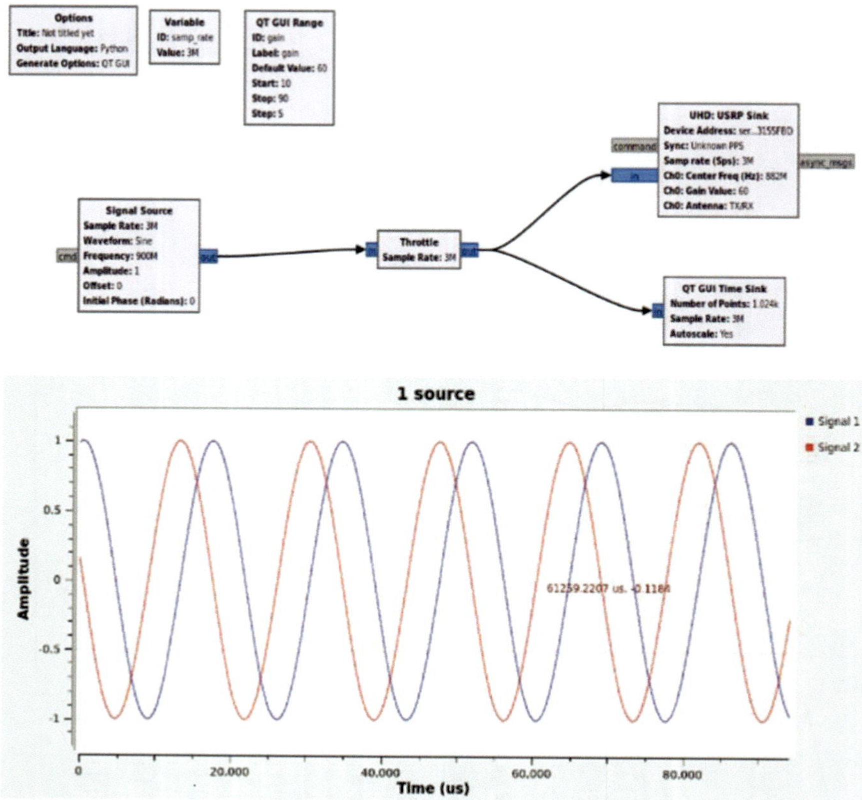

Figure 9.9 One transmitted signal GNU Radio flowgraph (top) and graph of signal.

the sample rate of 3,000,000. The signal is transmitted by the *UHD USRP sink block* with a gain of 60 db. Figure 9.9 shows the GNU Radio flowgraph of the transmitter and the plot of the transmitted signal.

Step 2 – time and phase synchronization:

To accurately receive the transmitted data, the USRPs in the ULA must be time-synchronized [30] before they even start receiving. The Octoclock (CDA-2990) synchronizes the time for the USRPs. All the receiving USRPs are connected to the Octoclock. A 10 MHz signal and a pulse per second (PPS) signal generated by the Octoclock are sent to each USRP in the ULA via cables for synchronizing the time across all the receiving USRPs. The connections can be seen in Figure 9.10.

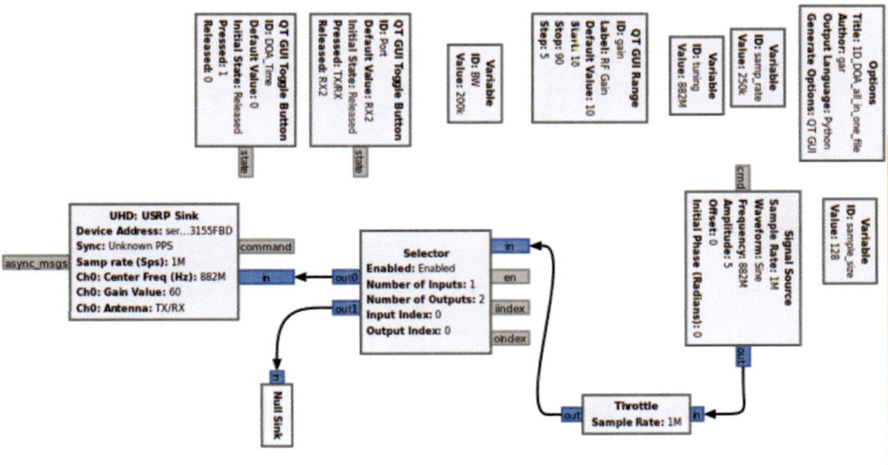

Figure 9.10 Reference signal phase correction section.

Phase correction is not a trivial task and has to be done each time before the DOA estimation process starts. The phase correction in our implementation is done through software in GNU Radio Companion. To correct the phase of the incoming signals transmitted, the phase difference/offset for each of the TX/RX channels on all the receiving USRPs is required. To find the phase difference/offset, a 10 MHz reference signal generated by a separate USRP is sent over coaxial cables to each of the RX2 channels on the receiving USRPs. The phase offsets calculated are then added to the transmitted signals received over the TX/RX ports. These steps are implemented in GNU Radio Companion in two stages as shown in Figures 9.10 and 9.11.

The *first stage* of the process is the synchronization/phase correction part of the flowgraph which controls the transmission of the reference signal and directs the signals received by the ULA to the phase correction/synchronization portion of the flowgraph. The synchronization portion of the flowgraph can be seen in Figure 9.11 below. To collect the phase difference, the incoming signal is first converted from complex to magnitude and phase components using the built-in *Complex to Mag Phase* block in GRC. The phase portion of the reference signal received by each antenna is then subtracted, using the *Subtract* block, by the phase of the reference signal received by the origin antenna. Then for testing purposes, the phase is corrected/synchronized by adding the phase differences to the incoming reference signal on each antenna other than the origin. The result of this step

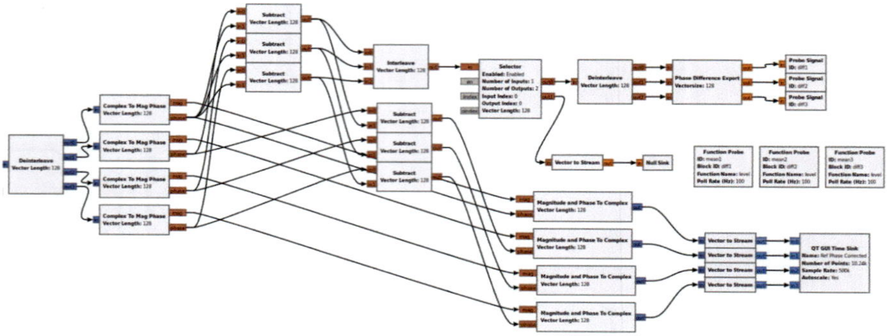

Figure 9.11 GNU Radio phase correction section.

is then displayed using one *QT GUI Time Sink* block that can be seen in Figure 9.11. At the same time, the phase differences are collected and stored for use in the DOA estimation portion of the flowgraph.

Figure 9.12 below shows the plots of the received signals before and after phase synchronization.

Step 3 – direction of arrival estimation:

The DOA estimation section begins with the phase-corrected signals being input into a *Python-embedded block* that runs the Python script of the proposed Capon-like method using LU decomposition, as shown in Figure 9.13. The DOA is estimated and outputted to the *number sink* and the *qt gui compass blocks*. The *number sink block* displays a numeric value on the user interface and the *qt gui compass block* displays the estimated angle on a compass with markings for 360°.

Two examples of the estimated DOA displayed on a compass and in the numeric form on GNU Radio's GUI can be seen in Figure 9.14. The first example has the source set at 15° and the second has it set at 40° from the receiving array.

9.5.4 Scenario 2: DOA estimation of two sources

Setup changes:

The setup for the second scenario with two sources remains the same except for the 2nd antenna connected to the transmitting USRP from the first scenario. Figure 9.15 shows the schematic diagram of the testbed setup for DOA estimation of two sources.

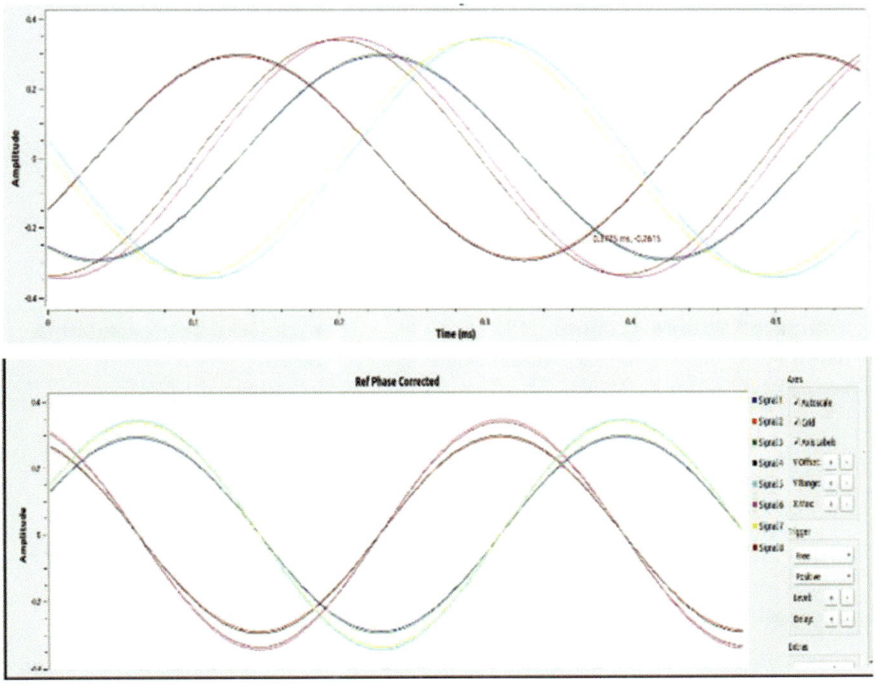

Figure 9.12 Signals before phase synchronization (top) and after (bottom).

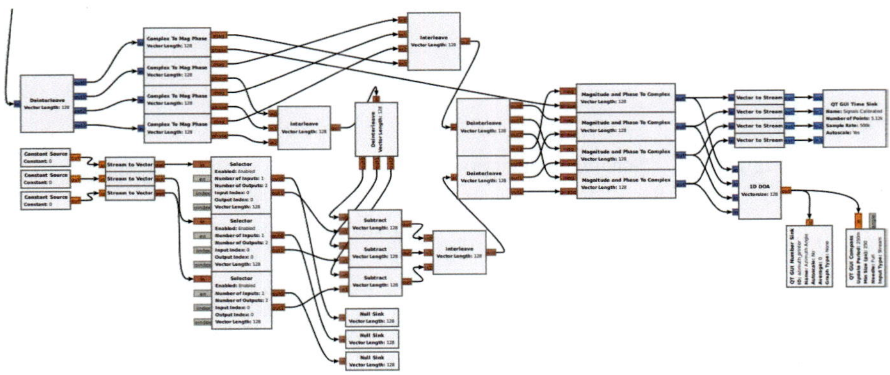

Figure 9.13 GNU Radio DOA section of the flowgraph.

Figure 9.16 below shows the two transmitting antennas in the testbed setup. Figure 9.18 shows the actual testbed built in the lab for the DOA estimation of two sources using NI USRP-2901 [29].

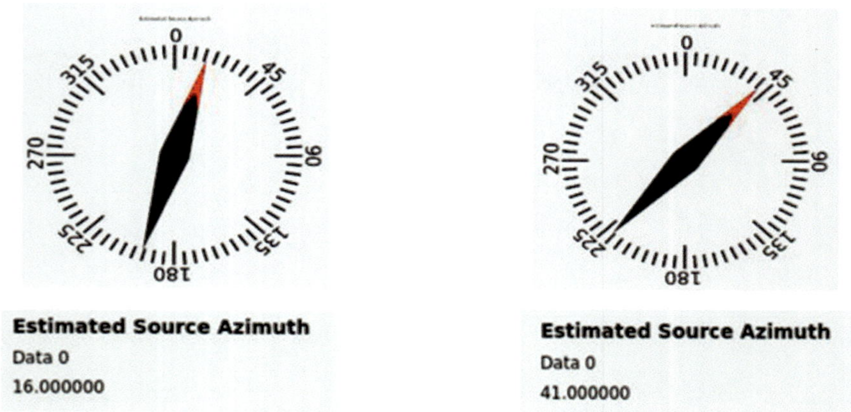

Figure 9.14 Example: source signal at 15° and 40°.

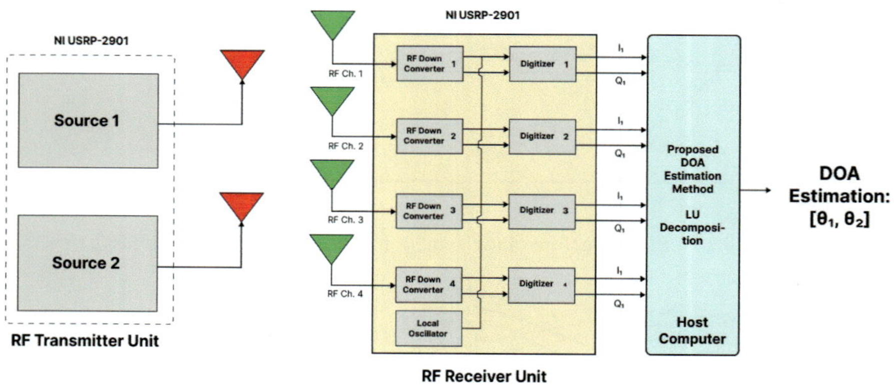

Figure 9.15 Schematic diagram of two sources DOA estimation setup.

Step 1 – Source signal generation:

As mentioned earlier, the *signal source block* is used to generate the two 882 MHz source signals at the sample rate of 3,000,000 and is transmitted by the *UHD USRP sink block* with a gain of 60 db. Diagrams for the flowgraph and the graphs of the transmitted signals can be seen in Figure 9.17 below.

Step 2 – time and phase synchronization:

No changes in the time and phase synchronization step are required for the DOA estimation for two sources.

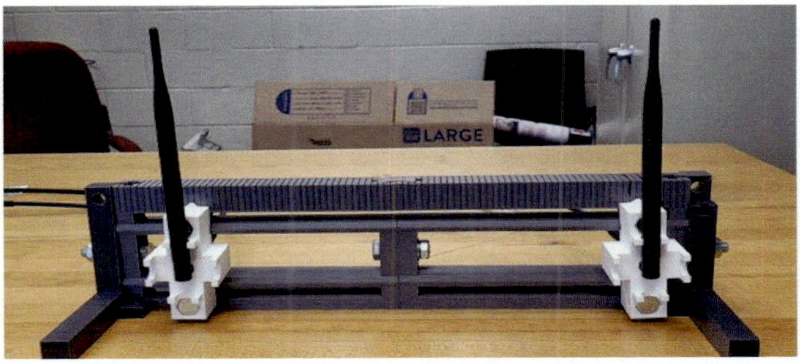

Figure 9.16 Two transmitting antennas.

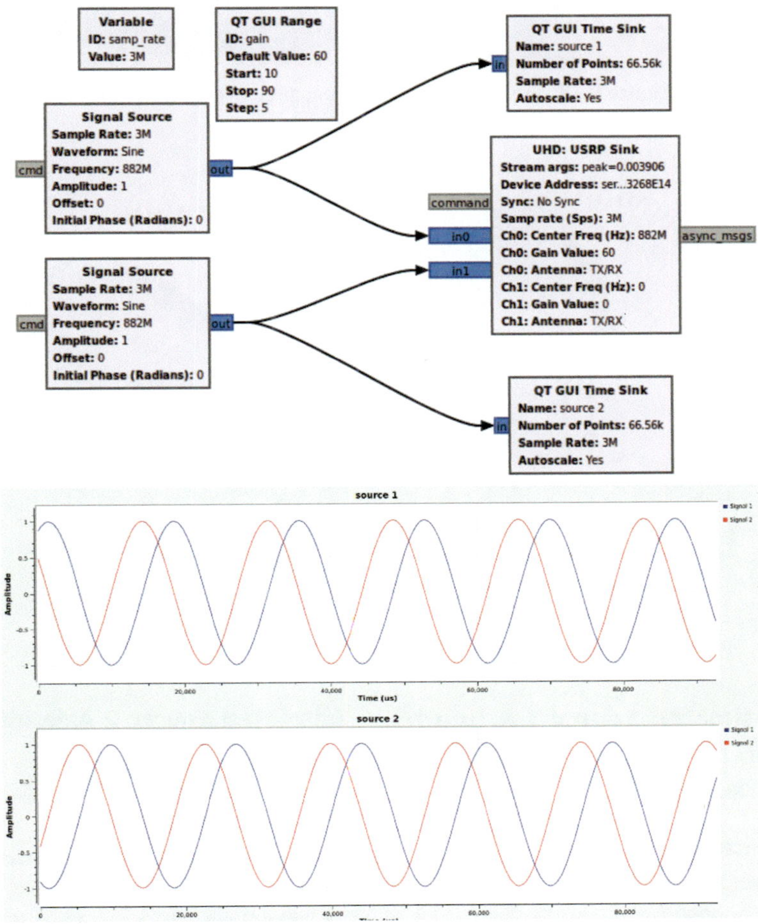

Figure 9.17 Two source transmit signals GNU Radio flowgraph (top) and signals plot.

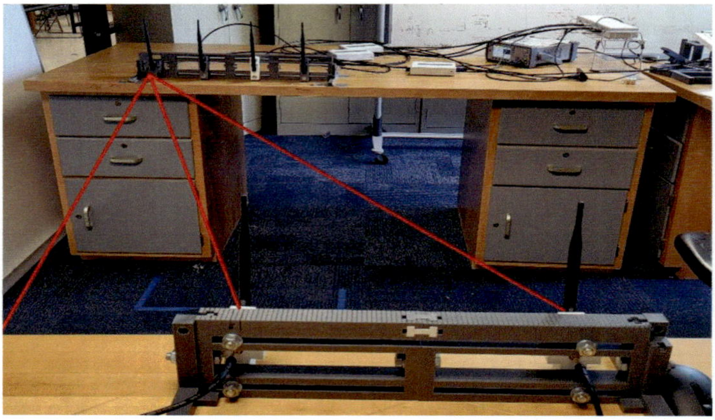

Figure 9.18 Two sources at different angles from each other.

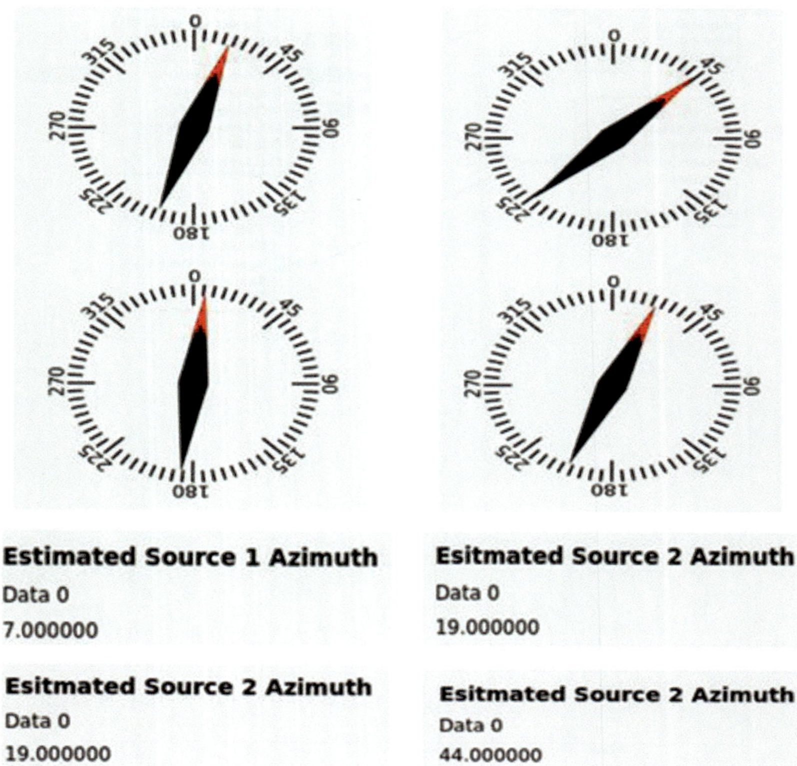

Figure 9.19 Example: source signals at 5° and 20°, and 20° and 45°.

Step 3 – direction of arrival estimation:

The direction of arrival estimation section also remains the same for the DOA estimation of two sources. Two examples of the estimated direction of arrival displayed on a compass and in the numeric form on GNURadio's GUI can be seen in Figure 9.19. In the first example, the sources are set at 5° and 20°, and in the second, they are set at 20° and 45° from the receiving antenna array.

9.6 Conclusion

This research chapter introduces a novel DOA estimation algorithm based on a combination of the Capon-like algorithm and LU decomposition, resulting in enhanced estimation accuracy while maintaining lower computational requirements compared to existing methods. Notably, it does not necessitate prior knowledge of the number of incident sources. To validate the effectiveness of the proposed algorithm, implementations were carried out in both MATLAB and GNU Radio flowgraphs. Additionally, a testbed utilizing a software-defined radio platform from National Instruments was utilized for real-time experimentation under diverse scenarios. The results obtained from MATLAB simulations and real-time testing demonstrate the proposed algorithm's capability to achieve accurate DOA estimation.

Acknowledgments

This research work was sponsored by the Air Force Research Laboratory (AFRL) in partnership with the Wright Brothers Institute (WBI) in the 5th year of a student challenge focused on Software Defined Radio (SDR) and Software Defined Networking (SDN) technology.

References

[1] Chen, Zhizhang, Gopal Gokeda, and Yiqiang Yu. Introduction to Direction-of-arrival Estimation. Artech House, pg. 9, 2010.

[2] Godara, Lal C. "Application of antenna arrays to mobile communications. II. Beam-forming and direction-of-arrival considerations." Proceedings of the IEEE 85, no. 8 (1997): 1195-1245.

[3] Compton Jr., R. T., Adaptive array - Concepts and Performance, Prentice-Hall, Englewood Cliffs, 1988.

[4] R.A. Monzingo, and T.W. Miller, Introduction to Adaptive Arrays. Wiley, New York, 1980.

[5] GNU Radio Companion - https://www.gnuradio.org/

[6] "What is GNU Radio? - GNU Radio." https://wiki.gnuradio.org/index.php/What_is_GNU_Radio%3F

[7] Gentilho, E., Scalassara, P.R. and Abrão, T., "Direction-of-arrival estimation methods: a performance-complexity tradeoff perspective," Journal of Signal Processing Systems 92 (2020):

[8] P. Gong and X. Chen, "Computationally Efficient Direction-of-Arrival Estimation Algorithms for a Cubic Coprime Array," Sensors, vol. 22, no. 1, p. 136, Dec. 2021, doi:10.3390/s22010136.

[9] Xu, Haiyun et al. "Two-Dimensional Direction-of-Arrival FastEstimation of Multiple Signals with Matrix Completion Theory inCoprime Planar Array." Sensors (Basel, Switzerland) vol. 18,6 1741.28 May. 2018, doi:10.3390/s18061741

[10] T. Ma, J. Du, and H. Shao, "A Nyström-Based Low-ComplexityAlgorithm with Improved Effective Array Aperture for Coherent DOAEstimation in Monostatic MIMO Radar," Remote Sensing, vol. 14, no.11, p. 2646, May 2022, doi:10.3390/rs14112646.

[11] Feng-Gang Yan, Zhi-Kun Chen, Ming Jian Sun, Yi Shen, and MingJin, "Two-Dimensional Direction-of-Arrivals Estimation Based on One-Dimensional Search Using Rank Deficiency", pp. 1-2, 2015.

[12] Roy, R., Kailath, T., "ESPRIT Estimation of Signal parameters via Rotational Invariance Techniques," IEEE Tran on Acoustics, Speech, and Signal Processing, vol. 29, no. 4, pp. 984-995, July 1989.

[13] P. Yang, F. Yang, and Z.-P. Nie, "DOA Estimation with Sub-array Divided Technique and Interpolated ESPRIT Algorithm on a Cylindrical Conformal Array Antenna," Progress In Electromagnetics Research, vol. 103, pp. 201–216, 2010.

[14] C. Qian, L. Huang, H. C. So, "Computationally efficient ESPRIT algorithm for direction-of-arrival estimation based on Nyström method," Signal Process., vol. 94, no. 1, pp. 74-80, 2014.

[15] Zhao Q., Liang W., "A Modified MUSIC Algorithm Based on Eigen Space," Advances in Computer Science, Intelligent System and Environment

[16] K. V. Rangarao and S. Venkatanarasimhan, "gold-MUSIC: A Variation on MUSIC to Accurately Determine Peaks of the Spectrum," in IEEE Transactions on Antennas and Propagation, vol. 61, no. 4, pp. 2263-2268, April 2013.

[17] Yaning Liu, Qin Zhang,Juntao Fu, and Yalei He, "A Semi-Real-Valued Capon Algorithm for 2-D DOA Estimation of Coherent Signals with L-Shape Array," 2018 2nd IEEE Adv. Information Management, Communicates, Electronic and Automation Control Conference.

[18] M. Bouri and S. Bourennane, "High resolution methods based on rank revealing triangular factorization," Transactions on Engineering, Computing and Technology, vol. 2, pp. 35-38, Dec 2004.

[19] Abdulrahman Alhamed, Nizar Tayem, Tariq Alshawi, Saleh Alshebeili, Abdullah Alsuwailem, Ahmed Hussain, "FPGA-based Real Time Implementation for Direction-of-Arrival Estimation," The Journal of Engineering, 2017, 13 pp., doi:10.1049/joe.2017.0165.

[20] Mulinde, R., Attygalle, M. and Aziz, S.M. (2021), Experimental validation of direction of arrival estimation for high chirp-rate linear frequency modulated radar signals. IET Radar Sonar Navig, 15: 627-640. https://doi.org/10.1049/rsn2.12069

[21] A. Hussain, N. Tayem and A. -H. Soliman, "LDL Decomposition-based FGPA Real-time Implementation of DOA Estimation," *2018 52nd Asilomar Conference on Signals, Systems, and Computers*, Pacific Grove, CA, USA, 2018, pp. 1163-1168, doi:10.1109/ACSSC.2018.8645387.

[22] Benahmed, A. & Mekhfioui, Mohcin & Hamzaoui, El-Mehdi & Guennoun, Zouhair. (2024). FPGA Implementation Of Adaptive Zero-Tracking Algorithm For Real-Time DOA Estimation Using A Linear Antenna Array. ASEAN Engineering Journal. 14. 207 - 212. 10.11113/aej.V14.20543.

[23] A. A. Hussain, N. Tayem and A. -H. Soliman, "Matrix Decomposition Methods for Efficient Hardware Implementation of DOA Estimation Algorithms: A Performance Comparison," 2019 4th International Conference and Workshops on Recent Advances and Innovations in Engineering (ICRAIE), Kedah, Malaysia, 2019, pp. 1-7, doi:10.1109/ICRA IE47735.2019.9037778.

[24] A. A. Hussain, N. Tayem, M. O. Butt, A. -H. Soliman, A. Alhamed and S. Alshebeili, "FPGA Hardware Implementation of DOA Estimation Algorithm Employing LU Decomposition," in IEEE Access, vol. 6, pp. 17666-17680, 2018, doi:10.1109/ACCESS.2018.2820122.

[25] Stoica, P., Händel, P. & Söderström, T., "Study of Capon method for array signal processing." Circuits Systems and Signal Process 14, 749–770 (1995). https://doi.org/10.1007/BF01204683

[26] P. Handel, P. Stoica and T. Soderstrom, "Capon method for doa estimation: accuracy and robustness aspects," IEEE Winter Workshop

on Nonlinear Digital Signal Processing, Tampere, Finland, 1993, pp. P_7.1-P_7.5, doi:10.1109/NDSP.1993.767766.

[27] Octo-clock CDA-2990 8-Channel Clock Distribution Module: https://www.ettus.com/product/details/OctoClock-G

[28] USRP SDR 2901: http://www.ni.com/en-lb/support/model.usrp-2901.html

[29] NI White paper - Building an Affordable 8x8 MIMO Testbed with NI USRP. April 2015. http://www.ni.com/white-paper/14311/en/

[30] Ettus Research Application Note – Synchronization and MIMO Capability with USRP Devices. https://kb.ettus.com/Synchronization_and_MIMO_Capability_with_USRP_Devices

10

Centralized Radio Access Network (C-RAN) Architecture for 5G using Fronthaul (FH) and Backhaul (BH) Networks

Hemanshi Anil K.[1], Vaibhav Gupta[1], Nandakumar S.[1], Sibaram Khara[2], and Velmurugan T.[1]

[1]School of Electronics Engineering, Vellore Institute of Technology, India
[2]Sharda University, India
E-mail: khemanshi.anil2015@vit.ac.in; vaibhav.gupta2015@vit.ac.in; snandakumar@vit.ac.in; sianba@rediffmail.com; tvelmurugan@vit.ac.in

Abstract

To keep up with the increasing demands of the next generation for high speed, low latency, high capacity, and pervasive mobile network, beyond 5G (B5G) evolution and beyond 5G features to predict the critical 6G requirements and highlight the 6G capabilities has been introduced in this research work. Centralized radio access network (C-RAN) architecture with respect to small cell architecture has been identified as a promising approach to address the benchmarks of 5G networks. To fulfill the unparalleled demands from users, the various rigorous needs for 5G networks with C-RAN on the basis of capacity, latency, and number of optical network units (ONU) supported and the cost has been compared. First, the capacity desideratum of the 5G fronthaul (FH) network has been approximated based on targeted data rates. Then, the requirements for fronthaul (FH) and backhaul (BH) networks were analyzed and compared the various scenarios to find the lowest latency. Further, for the calculated latency, the number of ONUs that can support our system has been found. The 5G C-RAN architecture has been simulated using R software and the throughout has been analyzed. A discrete event simulator has been used to analyze the results without any actual data set available. By tackling the complex problems associated with the rollout of B5G and

DOI: 10.1201/9788743801108-11

6G networks in rural areas, it hopes to promote a more equitable digital environment in which everyone can benefit from next-generation technology, regardless of where they live. Even if it is required that 6G make use of the greatest current network features and span a wider geographic area, we must first comprehend the significance of 6G.

Keywords: 5G, CRAN, crosshaul, CPRI, fronthaul, backhaul

10.1 Introduction

The next generation of mobile and machine-centric applications is driving increased needs for high capacity, low latency, and ubiquitous mobile access, which is driving the development of fifth-generation (5G) wireless technology. Utilizing the small-cell paradigm, the centralized radio access network (C-RAN) architecture is thought to be a viable way to address the requirements of 5G networks. However, a major difficulty in 5G C-RAN is guaranteeing end-to-end connection that is dependable, economical, and ensures quality of service. It is anticipated that the sixth generation (6G) mobile communication network will be essential in providing the necessary infrastructure, coverage, speed, and dependability of connections in the future. Because 6G devices can function in higher frequency bands than previous generations, they need a lot more energy than previous generations of devices. Energy efficiency and consumption are therefore serious problems that need to be resolved right away. Additionally, "Addressing B5G and 6G Network Connectivity Issues in Rural Regions" outlines a few tactical methods and fixes for these problems. It looks at public–private partnerships, financial sources, and legislative measures that can help with network infrastructure expansion. Widespread rural connectivity could be made possible by technological advancements like satellite communications, unmanned aerial vehicles (UAVs), and cutting-edge radio frequency technologies. The distribution and allocation of current mobile communications with 5G is about to happen for mobile network operators. The next sixth generation (6G) networking needs our full attention right now. We must first comprehend the significance of 6G, even if it is required to make use of the best current network features and cover a wider geographic area.

10.1.1 Motivation B5G and 6G

Next-generation mobile users crave faster data speeds and more reliable service. As people are getting mobile and machine-centric day by day, people

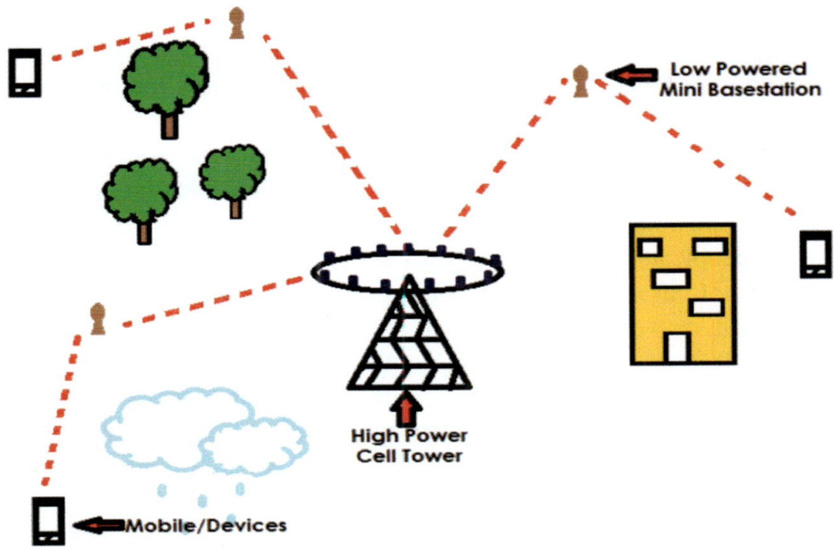

Figure 10.1 Small cell architecture.

and devices are consuming more data than ever before due to which the radio frequency spectrum less than 6 GHz (typically used by devices) is getting crowded and cramped up. To avert this problem, we use shorter mm waves that fall under 30 and 300 GHz. They are called millimeter (mm) waves because their length may vary from 1 to 10 mm, compared to the radio waves typically used in smartphones that measure tens of centimeters in length. But the problem is that mm waves cannot penetrate thick walls and buildings and may also get absorbed by trees and rain. To avoid this problem, we bring in small cells. As shown in Figure 10.1, small cells are mini base stations that can operate with lesser power consumption and can be placed every 250 m or so. This prevents dropped signals. Thousands of evenly spread base stations can form a relay to receive and transmit signals without interruptions. Small-cell deployment has become a vital way to improve coverage, capacity, and connection as wireless networks continue to develop. This chapter examines the idea of small cell technology, how it will be used in 5G and future 6G networks, as well as its advantages and disadvantages.

In addition to having more needs, 5G depends on technology that has not been widely used before. These astounding figures could lead one to believe that 5G is not only inevitable but that all future wireless communications objectives have been achieved.

Some debates regarding the subjects that will be left for future research are included in this chapter. Two steps comprise the presentation of these viewpoints. The first step is to pinpoint the 5G promises that are currently unattainable and for which technical and research collaboration is essential to developing workable solutions. The second will explore what lies beyond 5G, or what may be dubbed 6G; specifically, it will explore what uses are possible and what technologies may be crucial to enabling smooth communication between people and the cyber-physical world.

But as 5G users demand higher capacity and lower latencies, consider radio access networks (RAN), which consist of many stand-alone base stations (BTS). It consists of a remote radio head (RRH) and a baseband unit (BBU) at the cell site. To increase performance, enhance flexibility, and reduce cost in mobile fronthaul networks, use the next generation RAN, that is, centralized radio access network (C-RAN) [1–4]. In CRAN BBUs are relocated from individual cell sites to a centralized BBU. Removing BBUs from cells means operators can remove routers too. This simplifies networks and reduces the cost associated with space, heating, and cooling. Small cells need power sources and backhaul links that are effective. To guarantee high-capacity and sustainable networks in 6G, energy-efficient small-cell deployment combined with sophisticated power management and high-speed backhaul technologies will be essential.

Small cells can be swiftly deployed and positioned using unmanned aerial vehicles (UAVs) in locations that are hard to reach or where traditional infrastructure is impractical. Telecommunication service providers use UAVs to provide aerial network access in isolated rural areas, areas hit by disasters, or locations hosting large-scale events.

10.1.2 Technology trends and challenges

The main issue with the current state of mobile wireless system development is that to achieve the necessary throughput, densifying the network to the cells or expanding the spectrum is necessary. The need for mobile connectivity is only growing, and today's mobile and wireless communications cater to the anticipated multitude of devices and sensors found in machinery, automobiles, health systems, and urban infrastructures, in addition to the extremely dense populations of mobile phones and mobile computers. As promised by the fifth generation of radio communication networks, the notion of inclusive radio communication networks outlines the technologies for supporting wireless connectivity for all rates, kinds of communicating units, and scenarios. Considering the diverse range of wireless channels that will be used to provide this connectivity, as well as the heterogeneous nature of the future's highly interconnected human and machine systems. To transfer Qubits from one distant quantum processor to another, 6G will require quantum communication. By constructing quantum entangled Qubits, two quantum nodes can communicate with each other quantumly. Such a network can function as an intranet with a much fewer number of nodes.

The communication between unmanned aerial vehicles (UAVs) is a crucial aspect of multi-UAV systems design since it facilitates coordination and collaboration. Unmanned aerial vehicles (UAVs) can be employed in aerial sensor networks, which consist of several data sources dispersed over various regions, where UAV nodes are utilized to gather data. In situations where terrestrial infrastructure is compromised, UAVs present a possible possibility for sustaining cellular communication networks.

10.2 Requirements of 5G C-RAN

As depicted in Figure 10.2, it consists of a remote radio head (RRH) and a baseband unit (BBU). An RRH is a transceiver or antenna for the mobile network, whereas a BBU helps process calls and forward traffic. The RRH and BBU are interconnected with the help of the common public radio interface (CPRI) which can have a length of up to 15–20 km. A CPRI link transmits in-phase and quadrature signals (IQ) data of the baseband signals which requires a large bandwidth. A crosshaul network is a combination of fronthaul and backhaul networks. The fronthaul portion of C-RAN architecture comprises the intermediate link between the RRH and BBU, and the backhaul portion of it is the link between the core network and the small subnetworks at the edge network [5–8]. A 5G C-RAN architecture consists of three major components which are illustrated in Figure 10.2. RRHs that are deployed closer to the end users are connected to the centrally controlled BBU pool over the fronthaul network. Then, the BBU pools that are deployed in different geographical locations are connected through backhaul to the next level aggregation. The C-RAN architecture has been proven to be efficient in reducing CAPEX/OPEX, power consumption, and duration of the construction cycle.

The C-RAN approach advocates for the separation of the radio elements of the base station (called remote radio heads, RRHs) from the elements processing the baseband signal (called baseband units, BBUs), which are

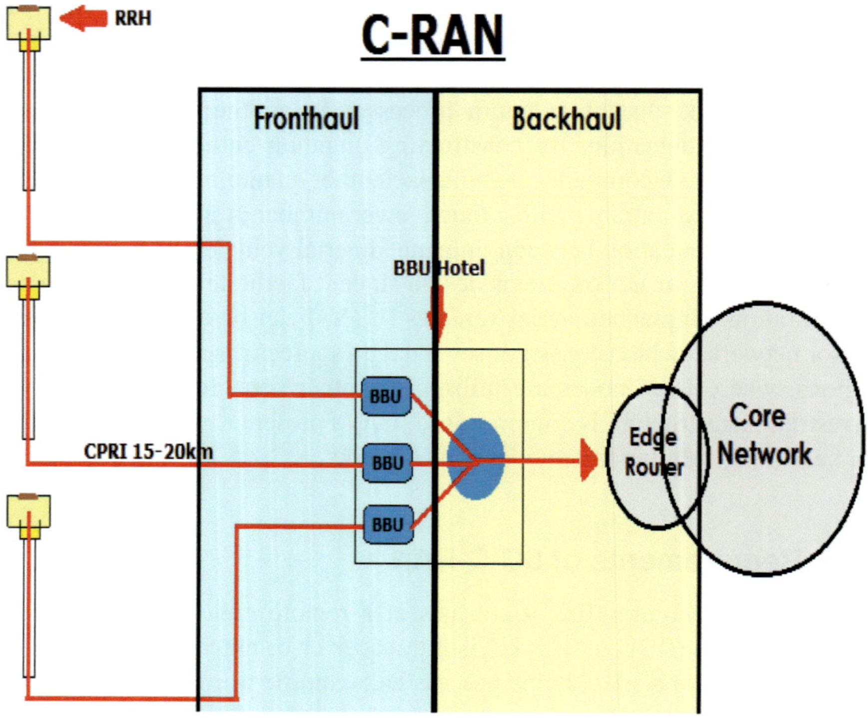

Figure 10.2 C-RAN architecture.

centralized in a single location or even virtualized into the cloud. RRHs that are deployed closer to the end users are connected to the centrally controlled BBU pool over the fronthaul network. Then, the BBU pools that are deployed in different geographical locations are connected through backhaul to the next level aggregation. This approach benefits from simpler radio equipment at the network edge, easier operation, and cheaper maintenance, while the main RAN intelligence (BBUs) is centralized in the operator-controlled premises. The cell contains both the remote radio head (RRH) and baseband units (BBU) in the other architectures. To reduce cost in mobile fronthaul networks, we use the next generation RAN, that is, centralized radio access network (C-RAN). In C-RAN architectures, all the baseband units are placed at a single site called a centralized BBU hotel. Removing BBUs from cells means operators can remove routers too. This simplifies networks and reduces the cost associated with space, heating, and cooling.

10.2.1 Capacity

To support high data rates and low latency, the 5G fronthaul network has high-capacity requirements. CPRI link transmits IQ data of the baseband signals. However, the transmission of IQ samples requires a larger bandwidth. For the radio to increase the bandwidth, we use multiple input multiple output (MIMO). To achieve a targeted data rate of 1 Gbps using 8×8 MIMO antennas across three sectors, a CPRI-based fronthaul network requires a total optical fiber link capacity of 147.5 Gbps. The required bandwidth of a CPRI link depends on the radio access technology in use, and it can be calculated using eqn (10.1).

$$C_{\mathrm{CPRI}} = S_{\mathrm{n}} \times A_{\mathrm{n}} \times F_{\mathrm{s}} \times BW_{\mathrm{s}} \times R_{\mathrm{c}} \times Ic, \tag{10.1}$$

where S_n is the number of sectors per RRH, A_n is the number of antennas corresponding to antenna configuration, F_s is the sampling frequency, BW_s is the sampling bit-width for I/Q samples, R_c is the ratio considered for the controlling overhead (a basic frame consists of 16 words and one used for controlling purposes), and I_c is the factor that accounts for the capacity increase due to 8B/10B encoding used.

10.2.2 Latency

In C-RAN, fronthaul (FH) traffic from the RRH has stringent delay requirements, whereas backhaul (BH) traffic from the BBU has mild delay requirements. The 5G framework design envisions a merged backhaul and fronthaul arrange which requires the fronthaul joins to move to an ethernet parcel-based arrange with pre-emption capabilities to guarantee bounded delays. In this study, we presented buffered frame pre-emption, stringent priority, and frame pre-emption with buffering strategies to enhance system performance and transport fronthaul and backhaul traffic streams simultaneously without compromising the quality of service. Generally, crosshaul packet forwarding elements (XPFEs) will forward both types of traffic, therefore introducing service differentiation which might improve the ability to fulfill the delivery guarantees of FH traffic, this is in line with the IEEE 802.1CM time-sensitive networking for fronthaul standard under development [9–12]. Here, we assume that XPFEs are at 75% load and operate at a 40 Gbps line rate, 50% of this load corresponds to FH traffic and the other 50% corresponds to BH traffic. CPRI (common public radio interface) with basic frames of 80-byte length, while BH packets follow the classical AMS-IX (Amsterdam

Internet Exchange) trimodal function, namely 7 out of 12 packets are short (40 bytes), 4 out of 12 are medium size (576 bytes) and 1 out of 12 packets is long (1500 bytes) [13].

10.2.2.1 Strict priority and frame pre-emption

A summary of the suggested methods such as strict priority and frame pre-emption employing buffering strategies to meet the QoS needs of both FB traffic and transmit the high-priority eCPRI fronthaul traffic and the low-priority backhaul traffic simultaneously. The highest priority packets are broadcast first from the queues in strict priority queuing regimes, and the lower priority packets are then communicated one at a time according to their priority settings. Priority traffic of lower priority cannot begin transmitting packets before all higher-priority packets from the queues have finished. Frame transmissions on legacy ethernet networks are supported in a nonpre-emption method, meaning that a frame in transit will always be completed and delivered to its destination without any interruption. Since a high-priority time-critical frame in this arrangement waits for a low-priority frame in the queue to finish transmitting, there are delays and variations in delays between high-priority traffic packets. We consider three different policies for service differentiation:

(1) No service differentiation.
(2) Strict priority (SP) is given to FH over BH, but without pre-emption.
(3) Strict priority is given to FH over BH with pre-emption.

The latency calculation flowchart is shown in Figure 10.3.

10.2.3 Number of ONUs

An optical network consists of an optical line terminal (OLT) which acts as a control switch module and decides what information floats over the optical distribution network (ODN), a splitter, and an optical network unit (ONU) that is located near the end user's premises as shown in Figure 10.4. We assume that a service operator is planning to install an antenna, carrying the mobile traffic over the ODN, and is considering two implementation options: deployment of a small cell to reduce the amount and the requirements of the generated traffic. Deployment of an RRH, following the C-RAN paradigm which would help generate time-sensitive FH traffic [14–15].

In the first case, we assume that the small cell generates 150 Mb/s peak traffic, and shares the ODN with seven residential users, which corresponds to

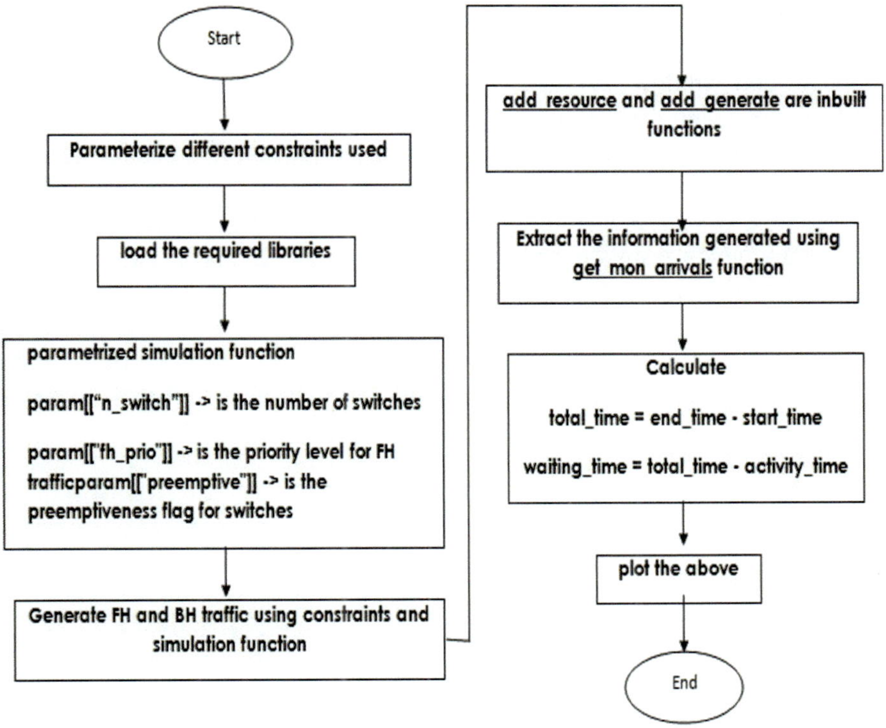

Figure 10.3 Flowchart of latency calculation process.

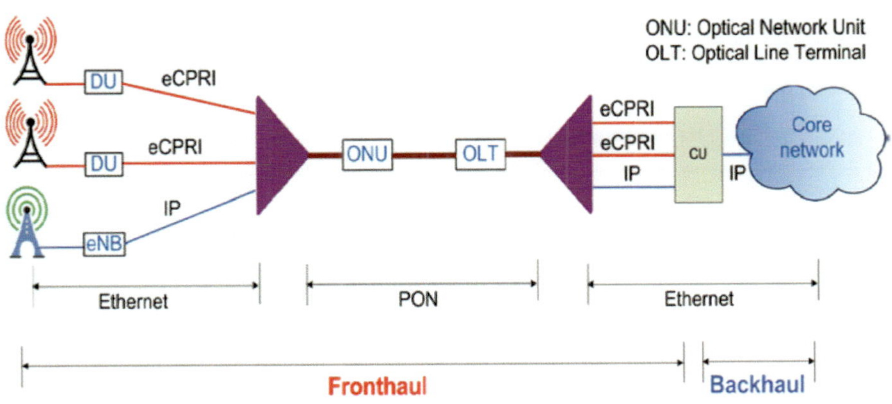

Figure 10.4 CRAN with ONU.

an average total load of 61.6%. In this case, we assume a dynamic bandwidth allocation (DBA) algorithm where each ONU requests at the end of its transmission window resources for the next cycle, while the OLT receives such requests and decides when and for how long each ONU may transmit its data in the next cycle granting transmission windows in a round-robin fashion. We assume a 1 μs guard time between transmission windows of consecutive ONUs, and in the second case, we assume an RRH generating FH traffic, which results in an approximate rate of 720 Mb/s. This FH traffic shares the ODN with now 31 residential ONUs, resulting in an approximate total load of 68.8%. To fulfill the tight delivery requirements of FH traffic, we assume that the DBA algorithm guarantees periodic TDM reservations for the FH traffic, and then the ONUs use a similar algorithm as before to share the

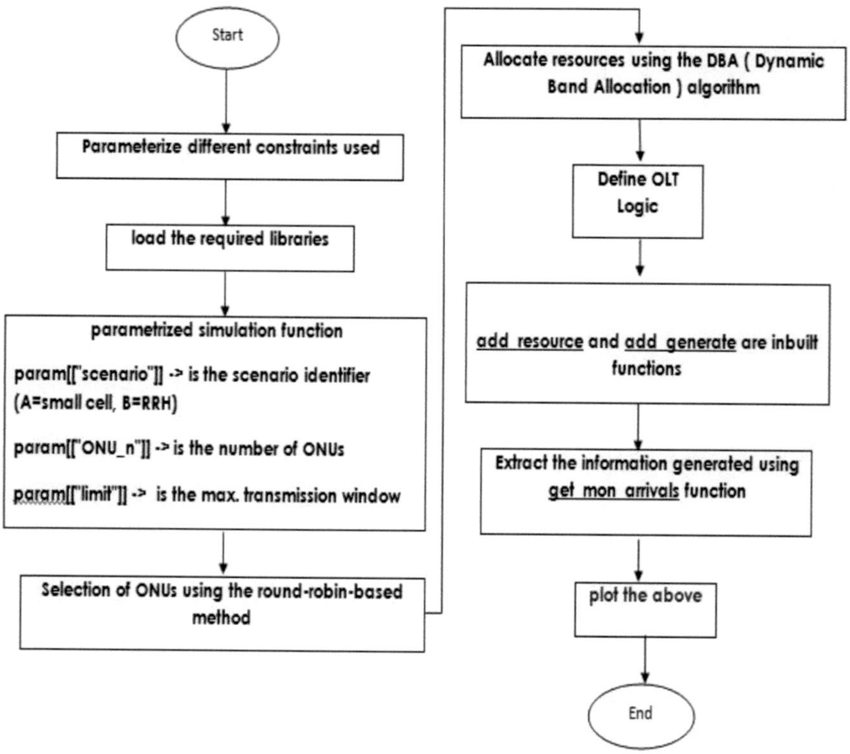

Figure 10.5 Flowchart of comparison of small cells and RRH with respect to ONUs.

rest of the available bandwidth. The implementation is shown in Figure 10.5. We have taken different numbers of bits namely 1500, 3000, 6000 and infinite bits to be allotted to the ONUs.

10.2.4 Cost

5G requires a very fast fronthaul architecture so that the user does not experience any kind of delay and also can cater to high bandwidth and lower delay budget. However, providing a proper fronthaul network in a cost-effective manner has become a challenge. Therefore, we equitably compare and analyze the deployment costs of C-RAN when it uses CPRI-based, physical layer split (PLS), and analog radio over fiber (ARoF) as a fronthaul network. To give a brief about PLS and ARoF, their architecture is represented in Figures 10.6 and 10.7, respectively.

We analyze several practical examples of our optimization framework by using the framework to plan a cost-optimal 5G and its optical fronthaul networks. The size of the area considered is consistent with a typical distribution area used in large-scale network planning. The capacity that can be supported by an RRH is assumed to be 1 Gbps. The normalized costs of RRHs and BBUs in the cases of PLS, ARoF, and CPRI are 220, 180, 200, 9500, and

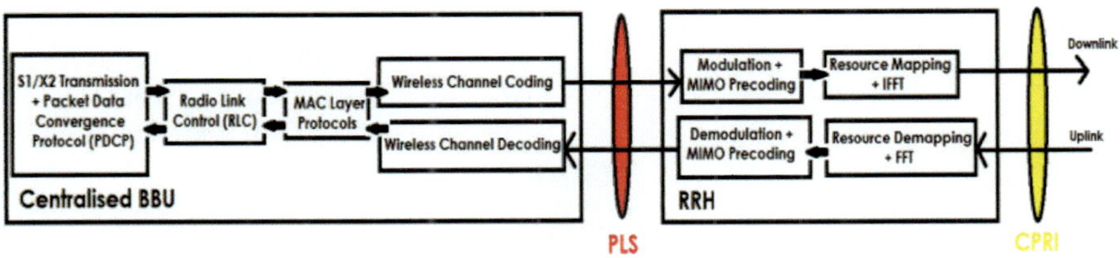

Figure 10.6 PLS architecture.

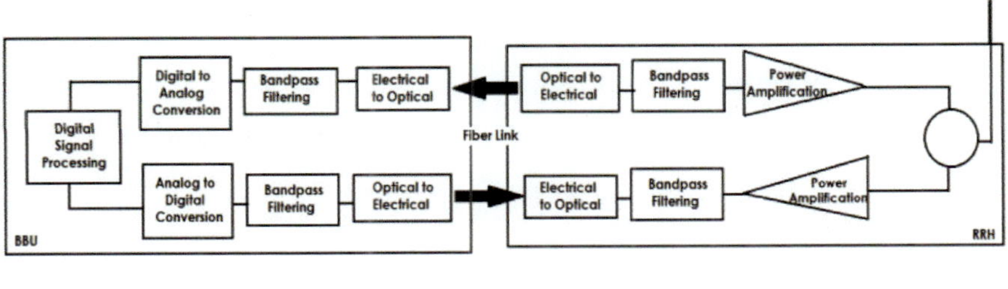

Figure 10.7 ARoF architecture.

10,500, respectively. For ARoF architecture, we use estimated cost values for RRHs and BBUs, due to the unavailability of accurate data. We consider a 5G deployment with carrier frequency less than 10 GHz when estimating these cost values.

10.3 Results and Discussions

10.3.1 Capacity

We evaluate the bandwidth requirement of a CPRI-based 5G fronthaul network by using the values listed in Table 10.1. Table 10.1 lists the typical values that can be used for these parameters in LTE and 5G networks. A CPRI link transmits IQ data of the baseband signals and it requires a larger bandwidth. The required bandwidth to support a CPRI link depends on the physical parameters of the radio access technology in use such as the number of MIMO antennas and wireless bandwidth [1]. To achieve the targeted data rate of 1 Gbps with 8×8 MIMO antennas from three sectors, 147.5 Gbps of optical fiber link is required for a CPRI-based fronthaul network. We can see that the requirement of capacity in 5G is way larger when compared to 4G as shown in Figure 10.8. The capacity requirement of CPRI links can be reduced

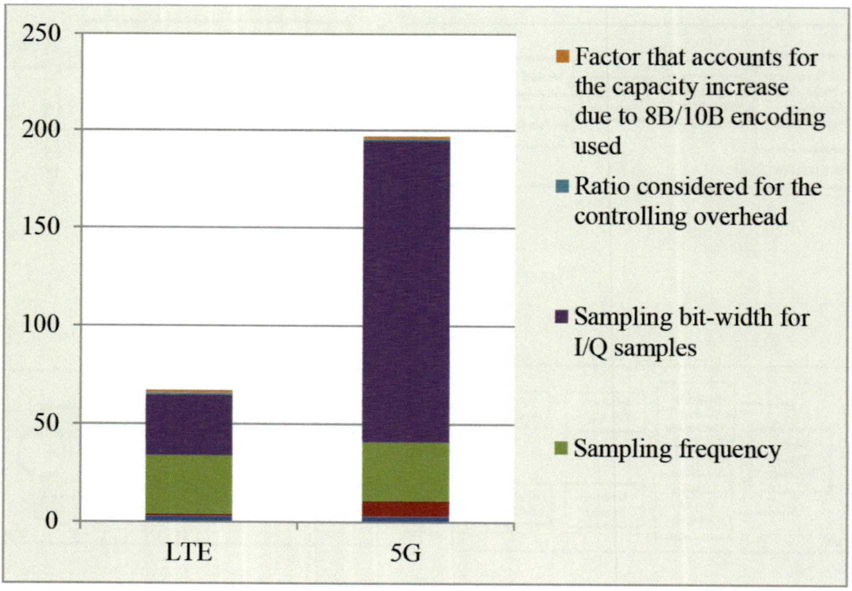

Figure 10.8 Comparison of 4G and 5G capacity requirements.

Table 10.1 Parameters for CPRI bandwidth calculation.

Parameters	**LTE**	**5G**
Number of sectors per RRH	3	3
Number of antennas correspond to antenna configuration	1	8
Sampling frequency	30	30
Sampling bit-width for I/Q samples	30.72	153.6
Ratio considered for the controlling overhead	1.06666	1.06666
Factor that accounts for the capacity increase due to 8B/10B encoding used	1.25	1.25

using compression techniques. However, despite these techniques can reduce the capacity requirement by up to 50%, still, a significant capacity is required by the CPRI link in Table 10.2.

10.3.2 Latency

The first approach ("without SP") results in both traffic experiencing the same queuing delay. The second approach ("with SP"), that is, service differentiation without pre-emption, results in a much-improved service for FH traffic, as FH packets only have to wait for other FH packets ahead and, sometimes, one BH packet being served. The third approach ("with SP and pre-emption") implements a pre-emption strategy, discarding BH traffic from the transmission line if an FH packet arrives, which, as shown in Figure 10.9, significantly improves delay performance which was caused by the long transmission times of long BH frames. For this scenario, we conclude that a pre-emption strategy significantly reduces queuing delay.

We analyze scenarios where the FH has to transverse N XPFEs in a cycle, each one also serving BH traffic. We depict in Figure 10.10 the queuing delays of FH traffic for $N = 1; 2; 5$, under the three considered policies (where the $N = 1$ case gives the same results that were computed in the previous experiments). As shown, FH queuing delay accumulates after traversing multiple XPFEs for the first and second agendas, both in terms of median and percentiles hence jitter too. Only the SP strategy with pre-emption keeps

Table 10.2 AMS-IX (Amsterdam Internet Exchange) trimodal function for BH packets.

CPRI (common public radio interface)				
Short	Medium	Long	Data rate	40Gb/s
7	4	1		
40 bytes	576 bytes	1500bytes	Load	0.75

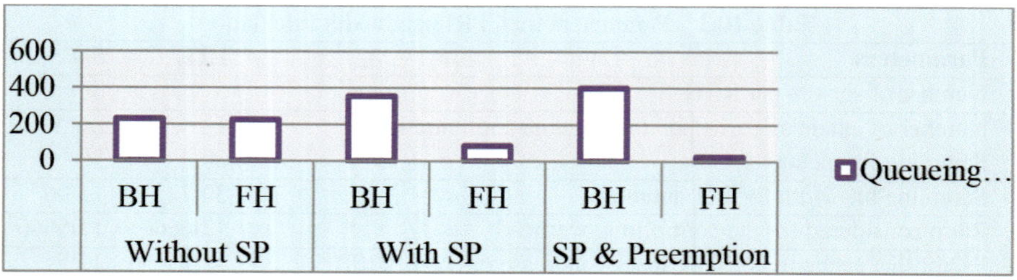

Figure 10.9 FH/BH queueing delay comparison for a single XPFE.

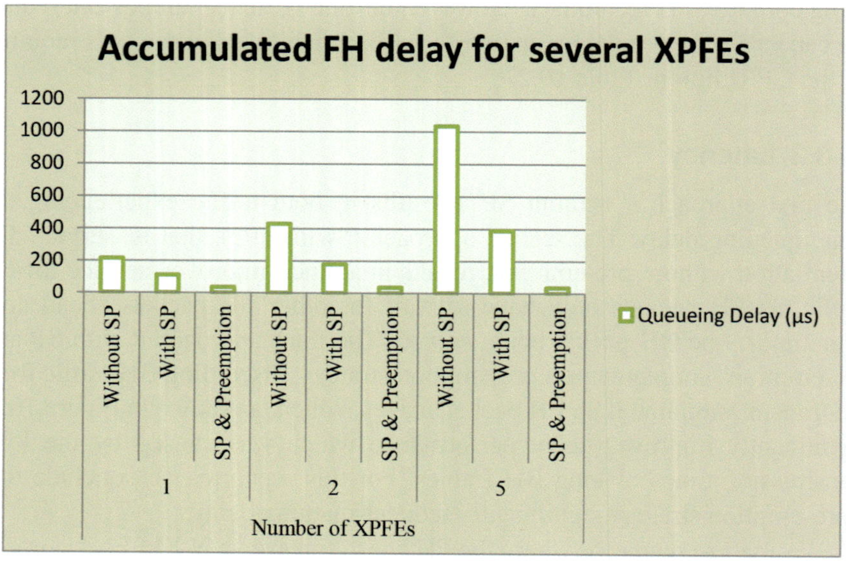

Figure 10.10 Accumulated FH delay for several XPFEs.

the delay constantly low throughout multiple XPFEs since FH packets only must wait if the server is occupied by other FH packets. In contrast, the SP strategy without pre-emption shows packet delay that keeps on increasing after traversing multiple XPFEs.

10.3.3 Number of ONUs

In the small cell case, as you can see, the delay varies quite a lot and is higher when there is no limit set whereas in the RRH case, we have only considered the infinite case as the delay remains constant throughout as seen above. In the RRH case, the delay is lower as compared to small cells while also

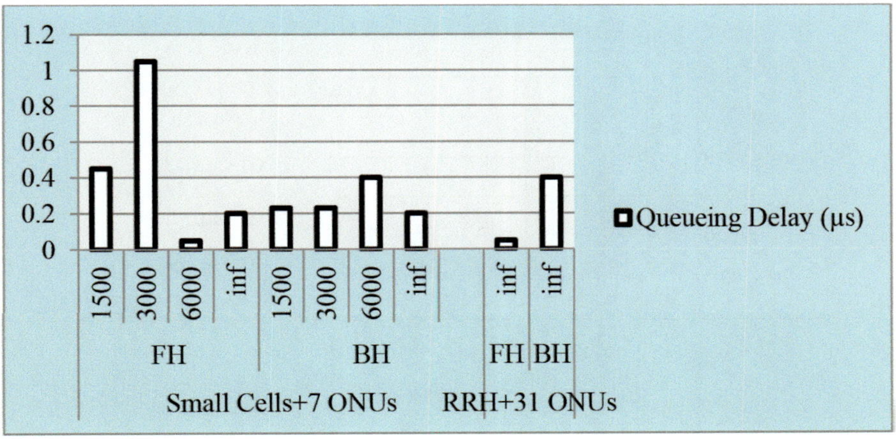

Figure 10.11 Comparison of small cells and RRH with respect to ONUs.

Table 10.3 Deployment of small cells versus RRH.

Small cells			**RRH**		
Peak traffic	Load	ONUs	Peak traffic	Load	ONUs
150 Mb/s	61.60%	7	720 Mb/s	68.80%	31

supporting a higher number of optical network units is shown in Table 10.3. The graphical representation is depicted in Figure 10.11.

10.3.4 Cost

Overall, the results show that to achieve cost-efficient network deployment, costs arising from both wireless access and optical crosshaul need to be considered during the planning stage along with other requirements as illustrated in Figure 10.12. Next, the total cost has been computed by considering both capacity and coverage requirements and it is shown in Figure 10.13. From Figures 10.12 and 10.13, it has been observed that to achieve cost-efficient network deployment, costs arising from both wireless access and optical crosshaul need to be considered during the planning stage along with other requirements. Conventional planning that considers stage-by-stage optimal network planning is inefficient when planning next-generation wireless access networks, where the fronthaul/backhaul network plays a critical role in determining the overall deployment cost. The results highlight the importance of employing a joint optimization framework in the planning of the next-generation wireless network along with its optical crosshaul counterpart.

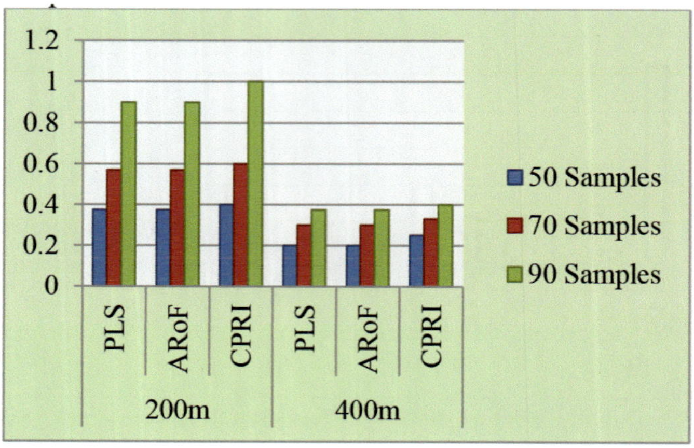

Figure 10.12 Normalized total cost versus population coverage requirement.

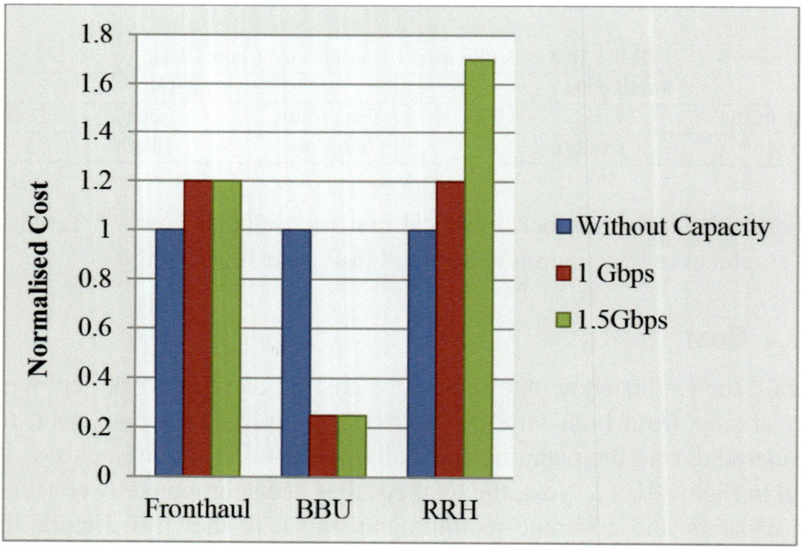

Figure 10.13 Optimal cost when both capacity and coverage requirements are considered.

The cost distribution of fronthaul, RRH, and BBU has been computed for various architectures and is illustrated in Figure 10.14. For example, if we carry out the optimal placement of 400 m radius RRH and PLS-based fronthaul network separately to achieve 90% of population coverage, there is a 15% cost increment compared to the joint optimization approach developed in this method [10].

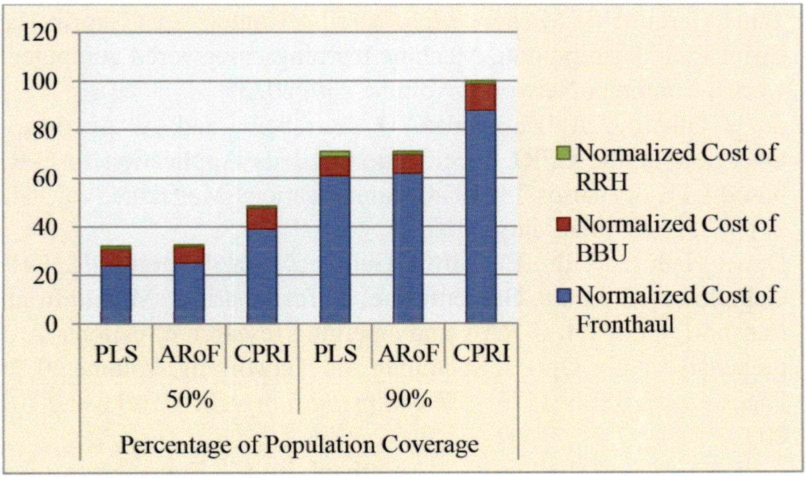

Figure 10.14 Cost distribution of fronthaul, RRH, AND BBU.

10.4 Summary

A key component of 5G networks, small cell deployment, allows for improved coverage, capacity, and connection. Small cells will be crucial in providing high data rates, low latency, and pervasive connection as the 6G network architecture takes shape. We have also investigated developmental aspects of the architecture like cost and capacity considering different fronthaul architectures as they contribute the most when considering the cost. CRAN architecture looks like one of the most efficient solutions for implementing 5G technology and this chapter gives an insight into some of the major requirements and challenges this architecture possesses. But 6G will bring with it new difficulties, like backhaul and power issues, network integration, and sophisticated antenna technology. Small cell deployment will revolutionize wireless connectivity as technology develops, opening the door to a seamless, fast, and dependable communication future.

References

[1] C. Ranaweera, E. Wong, A. Nirmalathas, C. Jayasundara and C. Lim, "5G C-RAN architecture: A comparison of multiple optical fronthaul networks," *2017 International Conference on Optical Network Design and Modeling (ONDM)*, Budapest, Hungary, 2017, pp. 1–6, doi:10.23919/ONDM.2017.7958544.

[2] Tania Cerquitelli, Michela Meo, Marilia Curado, Lea Skorin-Kapov, Eirini Eleni Tsiropoulou, Machine learning empowered computer networks, Computer Networks, Volume 230, 2023

[3] de la Oliva, J. A. Hernandez, D. Larrabeiti, and A. Azcorra, "An Overview of the CPRI Specification and its Application to C-RAN-based LTE Scenarios," IEEE Communications Magazine, vol. 54, no. 2, pp. 152–159, February 2016.

[4] Dawit Hadush Hailu, Berihu G. Gebrehaweria, Samrawit H. Kebede, Gebrehiwot G. Lema, Gebremichael T. Tesfamariam, Mobile fronthaul transport options in C-RAN and emerging research directions: A comprehensive study, Optical Switching and Networking, Volume 30, 2018, Pages 40–52, ISSN 1573–4277, https://doi.org/10.1016/j.osn.2018.06.003.

[5] A. Lometti and V. Sestito, "Fronthaul in 5G Transport Networks: IEEE1914.1 Architecture and Requirements," *2020 22nd International Conference on Transparent Optical Networks (ICTON)*, Bari, Italy, 2020, pp. 1–4, doi:10.1109/ICTON51198.2020.9203468.

[6] D. Wake et al, "Radio Over Fiber Link Design for Next Generation Wireless Systems,", Journal of Lightwave Technology, vol. 28, no. 16, pp. 2456–2464, 2010.

[7] F. Boccardi et al,"Five disruptive technology directions for 5G," IEEE Communications Magazine, vol. 52, no. 2, pp. 74–80, 2014.

[8] F. Malandrino, C. Casetti, C. F. Chias serini and G. Landi, "Optimization-in-the-Loop for Energy-Efficient 5G," *2018 IEEE 19th International Symposium on "A World of Wireless, Mobile and Multimedia Networks" (WoWMoM)*, Chania, Greece, 2018, pp. 1–9, doi: 10.1109/WoWMoM.2018.8449758.

[9] Md. Shahbaz Akhtar, Jitendra Gupta, Md. Iftekhar Alam, Sudhan Majhi, Aneek Adhya,Fronthaul latency and capacity constrained cost-effective and energy-efficient 5G C-RAN deployment, Optical Fiber Technology,Volume 80,2023,103392,ISSN 1068–5200,https://doi.org/10.1016/j.yofte.2023.103392.

[10] Mostafa Zaman Chowdhury, Takeo Fujii, Gabriel-Miro Muntean, Ji-Woong Choi, Giuseppe Araniti, "Emerging Small Cell Wireless Technologies for 5G: Architectures and Applications", Wireless Communications and Mobile Computing, vol. 2018, Article ID 8969264, 2 pages, 2018. https://doi.org/10.1155/2018/8969264

[11] Vuyo S. Pana, Oluwaseyi P. Babalola, Vipin Balyan, 5G radio access networks: A survey, Array, Volume 14, 2022, 100170,

[12] Alimi I.A., Teixeira A.L., Monteiro P.P. Toward an efficient C-RAN optical fronthaul for the future networks: A tutorial on technologies, requirements, challenges, and solutions IEEE Commun Surv Tutor, 20 (1)(2018), pp. 708–769

[13] J. S. Wey, Y. Luo and T. Pfeiffer, “5G Wireless Transport in a PON Context: An Overview,” in *IEEE Communications Standards Magazine*, vol. 4, no. 1, pp. 50–56, March 2020, doi:10.1109/MCOMSTD.001.19 00043

[14] Muhammad Waqar and Ajung Kim, QoS assurance for PON-based fronthaul and backhaul systems of 5G cloud radio access networks, International Journal of Internet Technology and Secured TransactionsVol. 11, No. 4, 2021, pp. pp 416–430.

[15] I. Ucar, J. A. Hernandez, P. Serrano and A. Azcorra, “Design and Analysis of 5G Scenarios with simmer: An R Package for Fast DES Prototyping,” in IEEE Communications Magazine, vol. 56, no. 11, pp. 145–151, November 2018, doi:10.1109/MCOM.2018.1700960.

11

A Dynamic Radio Resource Allocation Scheme for the B5G Mobile Systems Evaluated on the B5G/IoT Testbed with High Reliability and High Elasticity

F. Kojima[1], R. Nakano[1], and T. Ikenaga[2]

[1]Social Innovation Unit, Open Innovation Platform Headquarters, National Institute of Information and Communications Technology (NICT), Japan
[2]Department of Electrical and Electronic Engineering, Faculty of Engineering, Kyushu Institute of Technology, Japan
E-mail: f-kojima@nict.go.jp; n.renri@nict.go.jp; ike@ecs.kyutech.ac.jp

Abstract

This chapter presents a dynamic radio resource allocation scheme according to the radio utility conditions for B5G mobile applications. This scheme is evaluated in a highly reliable and highly elastic B5G/IoT testbed that provides wired and wireless network evaluation environments with adequate backup infrastructure and expandable functions via software-defined technologies. The testbed is implemented by integrating the conventional Japan Gigabit Network (JGN) wired network and the StarBED computing environment, and mobile radio area facilities customized from the Local 5G (L5G) system that is a private 5G system by restricting the system bandwidth and system deployment from the original systems. The proposed dynamic resource allocation scheme is implemented through a software-defined mobile base station that supports software-based configurations for central unit (CU), distributed unit (DU), and core network (CN). Additionally, the proposed scheme incorporates remote and distributed core facility deployment, enhancing system

DOI: 10.1201/9788743801108-12

orchestration capabilities. The performance of the dynamic resource allocation scheme is validated through experiments on the testbed, demonstrating its effectiveness. The study further explores advanced B5G mobile application scenarios with multiple base station deployments. Results indicate a 120% improvement in normalized radio resource utilization across multiple base stations when using a slow adaptive modulation technique. This improvement is observed with a normalized cell radius parameter (which determines the relatively higher gain modulation employed in the central cell region) $R_{sams} = 0.5$ and a normalized gain parameter (which determines the gain of such employed modulation) $G_{sams} = 2.0$, compared to scenarios without adaptive modulation, that is, $R_{sams} = 0.0$ and $G_{sams} = 1.0$.

Keywords: Testbed, B5G, mobile, radio resource, experiment

11.1 Introduction

The radio systems in beyond 5th generation (B5G) or 6th generation (6G) networks are evolving to meet increasingly complex and diverse service demands driven by advancements in information and communication technology (ICT) and the support of cyber-physical systems (CPS) [1]. This progress includes wireless emulation technology, which evaluates wireless systems or services by analyzing radio wave performance in cyberspace without actual radio emissions [2–4]. To accurately assess the performance of these diversified systems, testbed environments are essential. A testbed provides an isolated evaluation environment for future systems or services, offering necessary facilities or content for testing without impacting existing systems.

The ICT Testbed Research and Development Promotion Center at the National Institute of Information and Communications Technology (NICT) has been actively involved in creating and providing such testbed environments, known as the "NICT ICT testbed." Previously, this testbed mainly featured high-speed wired network infrastructure, such as the Japan Gigabit Network (JGN) [5], and a large-scale computation platform, StarBED [6]. However, with the evolving needs of future systems, expanding the functionality of the testbed has become imperative. Since October 2022, NICT has launched the B5G/IoT testbed, offering enhanced reliability and flexibility. This testbed integrates previous capabilities and introduces new evaluation features tailored for B5G systems and services.

The expanded testbed environment facilitates comprehensive evaluations across different layers: network layer evaluations on both wired and wireless infrastructures, middleware evaluations using simulation and emulation technologies with visualization features, and platform layer evaluations incorporating big data analysis and application development. Notably, the wireless communication evaluation environment of the testbed, designed for B5G mobile applications, is crucial for advancing new radio systems and services that utilize effective radio resource management. While several studies have explored B5G system evaluations using testbed environments [7–9], they have not extensively considered effective radio resource management. This chapter addresses this gap by proposing a dynamic radio resource allocation scheme for B5G mobile systems. The scheme involves software-based system customization in local distributed units (DU), central units (CU), and the core network (CN). Additionally, the chapter examines the deployment of multiple base stations using slow adaptive modulation technology [10], which enhances spatial radio resource reuse compared to traditional temporal-focused adaptive modulation methods [11].

The key contributions of this chapter are as follows:

1. It provides an overview of the developed B5G/IoT testbed, highlighting its high reliability and flexibility, as well as its ability to support evaluations across network, middleware, and platform layers through software-defined technologies.
2. It proposes an effective radio resource allocation scheme adaptable to varying system conditions during both normal and emergency situations, such as disasters, and evaluates its performance using the enhanced B5G/IoT testbed.
3. It demonstrates the potential for significant functional enhancements by incorporating slow adaptive modulation technology.

The remainder of this chapter is organized as follows: Section 11.2 outlines the initiatives of the NICT in testbed development and their alignment with future B5G system requirements. Section 11.3 describes the detailed construction and functionalities of the high-reliability, high-flexibility B5G/IoT testbed, including aspects of the B5G mobile environment. Section 11.4 explains the proposed radio resource allocation scheme and presents performance evaluation results obtained by using the testbed. Section 11.5 discusses the potential functional expansions enabled by slow adaptive modulation technology. Finally, Section 11.6 concludes the chapter.

11.2 Testbed Provision Activities for the Enhanced and Diversified Systems and Services in B5G Society

11.2.1 An expectation of the B5G system requirements

Figure 11.1 shows a schematic of expected B5G system requirements accompanied by that for 5G and before systems. As shown in Figure 11.1, 4G and earlier systems assume a single requirement, such as a high data rate,

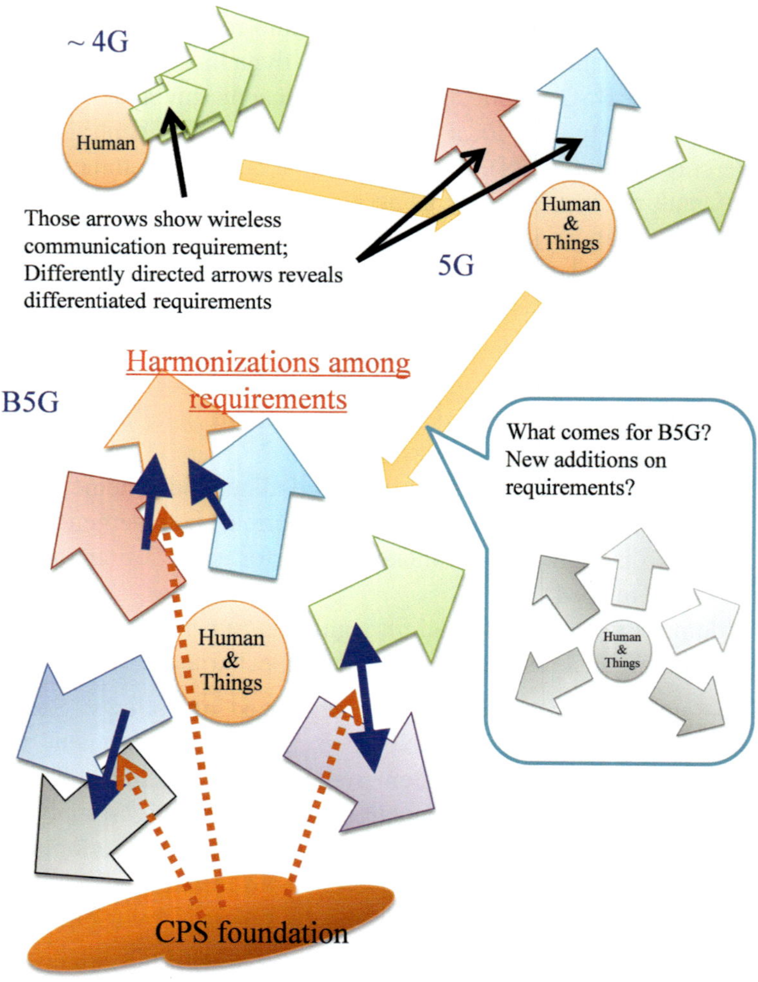

Figure 11.1 An expectation on B5G system requirements assuming harmonizations conducted by the CPS foundation.

and mainly consider the radio system utilization by humans. In 5G systems, on the other hand, diversified radio communication system functions enable applications to IoT systems, as well as human systems accompanied by the increase of the vertical sectors. Here, multiple requirements, such as high data rate, low latency, and massive connectivity, are expected according to the system utilization conditions.

In the case of the B5G system, one possible expectation might be a simple addition to the multiple requirements. However, in Figure 11.1, further enhancements are expected and presented; that is, harmonization among such multiple requirements is considered in this B5G system. The harmonization can include combination, negotiation, and compromise that leads to further diversified and complicated requirements. In order to suitably conduct those harmonizations, a management function by the CPS foundation is considered necessary as in Figure 11.1. It is needless to say that the testbed environment for the evaluation of B5G systems and services needs to support those harmonized requirements.

11.2.2 Testbed development directions deploying the network, middleware, and platform layers

Figure 11.2 shows a direction of testbed development and provision with consideration of enhanced evaluation capability assuming the harmonized requirements in Figure 11.1. In order to realize the suitable evaluation environment, not only the evaluation environment on the network layer that provides communication infrastructures as in NICT's conventional testbed

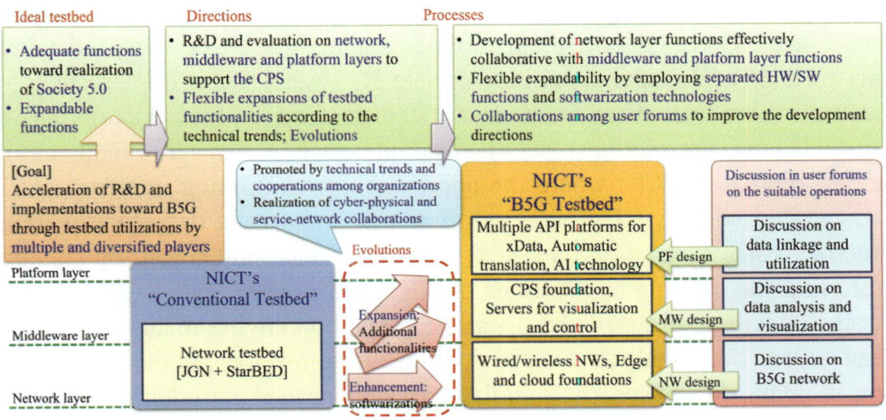

Figure 11.2 Testbed enhancement and provision directions considered in NICT.

that provides JGN and StarBED environments but also that on the middleware layer and platform layer becomes essential factors.

Furthermore, functional enhancements in the conventional evaluation environment are also required to suitably cope with the advanced system performances. System customization technologies employing software implementations are considered very effective in such enhancements.

11.3 B5G/IoT Testbed with High Reliability and High Flexibility Provided by NICT

11.3.1 Overall testbed structure that employs environments to evaluate B5G technologies in each evaluation layer

In order to realize a suitable evaluation environment for B5G systems and services, NICT has developed a B5G/IoT testbed with high reliability and high flexibility and started its provision activities in October 2022. Compared with the NICT's previous testbeds of JGN and StarBED, the proposed testbed has a newly deployed "B5G Reliable Virtualization Environment" that provides an evaluation environment utilizing hardware resources extensively and prevents network service failures and newly achieving reliability. Moreover, "B5G Reliable Virtualization Environment" and "B5G Mobile Environment" can also enable dynamic and flexible communication resource management, which newly provides flexibility that was not supported in the previous testbeds. Figure 11.3 shows the overall construction of this testbed environment. Here, the network layer of the testbed is enhanced to realize the software infrastructure testbed where mobile, optical, and virtualization environments will be provided with flexibility and reliability by deploying the Local 5G (L5G) compliant mobile systems and JGN advanced wired network [12]. The L5G system is a private 5G system that is implemented by restricting the system bandwidth and system deployment from the original 5G systems [13]. The platform layer testbed introduces a concept of data-centric cloud service (DCCS) where the application programming interface (API) cooperated with cloud service to create new values via data coordination, such as voice translation data, text translation data, and so on, by suitably referring the existing applications or NICT developing applications. DCCS testbed environment provides a service evaluation environment where users can utilize data analysis functions, as well as develop applications based on the data. Moreover, the middleware layer testbed introduces emulation infrastructure coupled with the real world called CyReal where simulations,

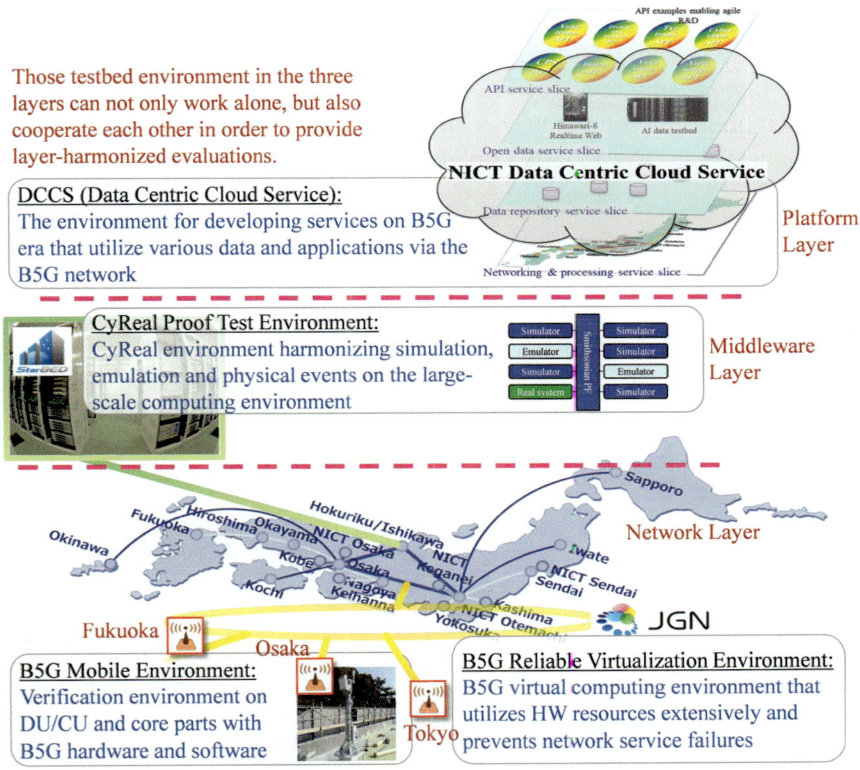

Figure 11.3 The overall structure of the B5G/IoT testbed with high reliability and high flexibility provided by NICT.

emulations, and real systems can be linked through common interfaces to develop new services in the StarBED-oriented computing environment [14]. One of the typical examples of evaluation using CyReal environment is the evaluation of the disaster situation that employs flooding simulation, disaster prevention radio emulators, and the actual monitoring and visualizing systems. The testbed environment in the three layers can not only work alone but also cooperate with each other in order to provide layer-harmonized evaluations.

11.3.2 Details of the B5G mobile environment that employs evaluation environments of the proposed scheme

In this subsection, details of the B5G mobile environment that is included as an evaluation environment on the network layer in B5G/IoT testbed with high reliability and high flexibility that provides wireless communication evaluation assuming B5G mobile systems and is related to the proposed dynamic radio resource allocation scheme. Figure 11.4 shows an outline of the B5G mobile environment. This environment includes three environments as in Figure 11.4. The mobile application evaluation environment employs

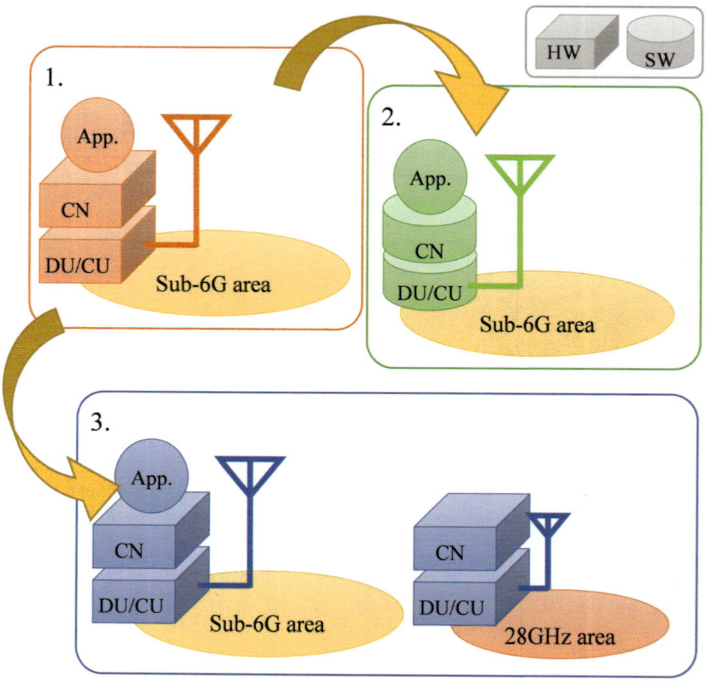

1. Mobile application evaluation environment:
HW base stations (BSs) for evaluations of application technologies exploiting 5G network assuming B5G.

2. Mobile network development environment:
Cloud native BSs that enables performance evaluations for BS customizations such as C-Plane slicing technology.

3. Mobile base station development environment:
Heterogeneous BSs (28 GHz, Sub-6 GHz) and multi-band terminals to evaluate heterogeneous area managements.

Figure 11.4 Outline of the B5G mobile environment.

hardware (HW) base stations (BSs) that enable evaluations of application technologies by assuming the application software installation and exploiting the 5G network assuming enhancement toward B5G. The mobile network development environment employs cloud-native BSs to enable performance evaluations for BS customizations by softwarized technologies such as C-Plane slicing, as reported in the following sections. Therefore, software-based DU, CU, and CN customization is conducted only on the mobile network

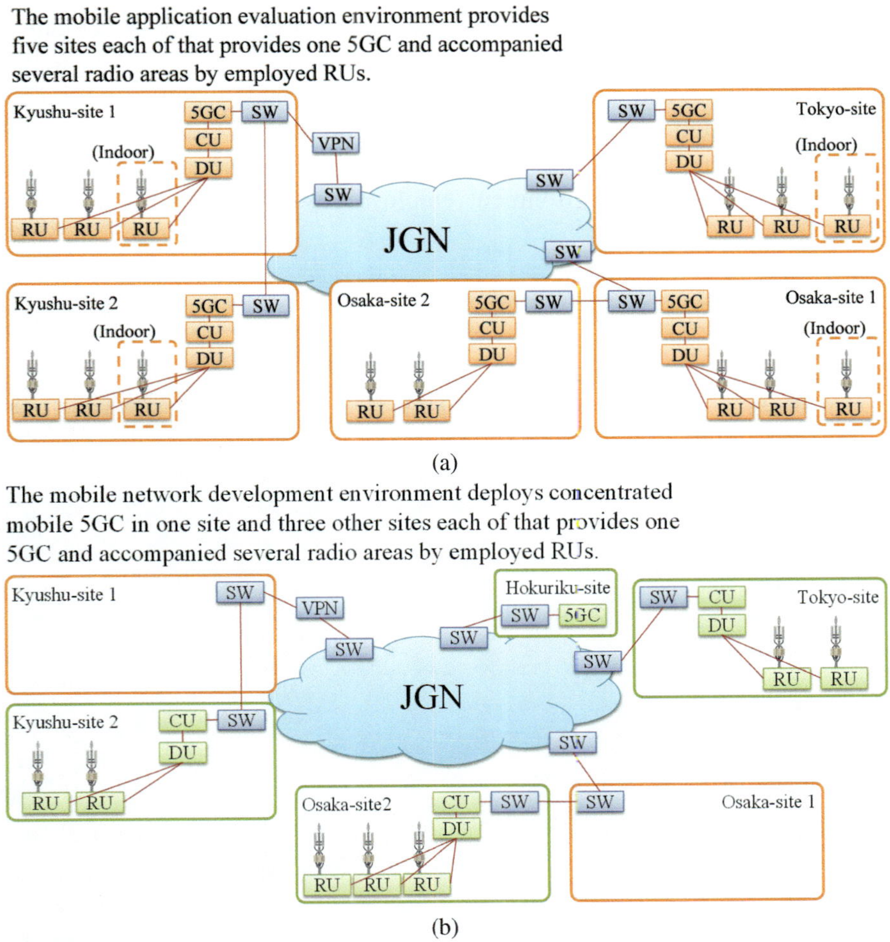

Figure 11.5 Outline of mobile evaluation areas and equipment deployed in the B5G mobile environment. (a) Deployment for the mobile application evaluation environment. (b) Deployment for the mobile network development environment.

development environment. The mobile base station development environment provides heterogeneous BSs employing 28 GHz, as well as sub-6 GHz, and multiband mobile terminals; thereby, users can evaluate the hybrid radio area utilization situations.

Figure 11.5 shows the deployment of the mobile application evaluation environment and the mobile network development environment. Those environments assume distributed sites via JGN-supported connection as in Figure 11.5. One of the BS facilities constructed in the Tokyo-site in the NICT headquarters is shown in Figure 11.6. As depicted in Figure 11.5, the mobile application evaluation environment and the mobile network development environment deploy their own BSs partially in the same sites such as Tokyo-site, Osaka-site2, and Kyushu-site2. However, the most typical difference is that CU, DU, and CN customization is enabled only in the mobile network development environment as mentioned previously. Therefore, the mobile network development environment employs customizable CNs as remote CNs.

Figure 11.6 A BS facility constructed on the Tokyo-site.

11.4 Proposed Dynamic Resource Allocation Schemes Evaluated in the Testbed Environment

11.4.1 Details of the proposed resource allocation scheme assuming software-based system customizations

This section reports on the proposed dynamic radio resource allocation scheme that is evaluated in the B5G mobile environment. Figure 11.7 shows outlines of the proposed scheme. The proposed scheme includes two main managements in order to realize effective radio resource utilization. One is

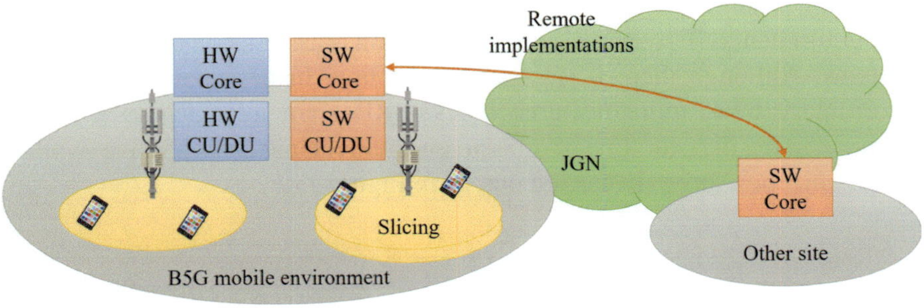

(a)

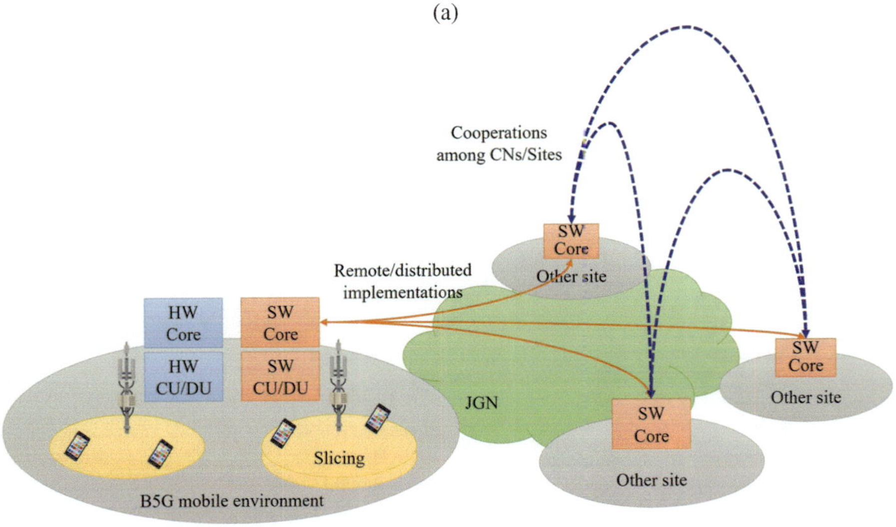

(b)

Figure 11.7 Outlines of the proposed scheme. (a) A concept of resource allocation by slicing. (b) Further management by the core configurations.

the dynamic resource allocation by the slicing technology in CU/DU units. The other is a network configuration that provides remote and distributed CNs, as shown in Figure 11.7. In this management, the software-based CNs, as depicted in Figure 11.5 (b), can be deployed as remote and/or distributed CNs that provide further flexible and complicated management for their mobile users. Both of those two managements effectively utilize the software-based system customizations assumed in the mobile network development environment. The former can be implemented by conducting a slicing setting in the CU/DU. The latter can be implemented by deploying software-based CNs in remote or distributed sites.

Figure 11.8 shows a typical application of the proposed radio resource allocation scheme that is deployed for disaster management activities. Several user equipment (UE) terminals accommodation is assumed in the B5G mobile systems. Although there is no specific resource management is employed in ordinary time, more resource blocks (RBs) are allocated to a certain terminal in the disaster time in order to ensure dynamic and effective wireless communication links are carried out, as in Figure 11.8.

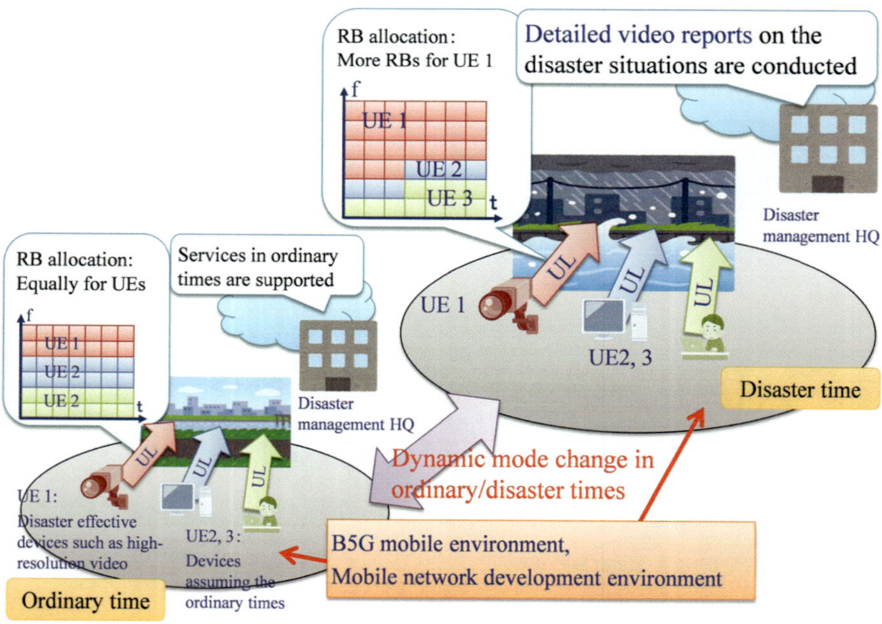

Figure 11.8 A typical application of the proposed radio resource allocation scheme to disaster management activities.

11.4.2 Experimental evaluation of the proposed system on the mobile network development evaluation

In this subsection, an experimental evaluation of the proposed radio resource allocation scheme conducted on the mobile network development environment is presented. Especially, in this chapter, mobile applications performance provided by the remote and distributed CNs is presented. The conditions of the experiments are shown in Table 11.1. Those conditions are compliant with the current Local 5G (L5G) systems in Japan, which are put into practical use for the 5G-based special local radio communication system and allocated in the 4800 to 4900 MHz band. The employed frequency and bandwidth mean a center frequency and a bandwidth defined in this L5G specification, respectively.

In the experiments, performances of downlink (DL) throughput and round-trip time (RTT) are obtained in the mobile network development environment in Tokyo-site where a 5-m radio propagation distance is employed between a UE and a BS. Figure 11.9 shows observed performances on the developed visualization applications that provide performance comparisons between cases with different CN deployments on Hokuriku-site and Tokyo-site, as depicted in Figure 11.5. Median values for both performances are summarized in Table 11.2. Although the CN in Tokyo-site means the experimental situation where the mobile service area with a BS and a terminal is located in the same site as NICT headquarters, the CN in Hokuriku-site is apart from the mobile service area by a straight-line distance of about 300 km. Therefore, as in Table 11.2, RTT performance degradation is confirmed in the case of CN in Hokuriku-cite. Moreover, DL throughput performance is also degraded compared with CN in Tokyo-site, providing equivalent performances in the experiments in [9] employing CN in the same site. This is because such transmission latency performances, as in RTT evaluations, also prevent effective RB allocation utilizing the preliminary employed RB set. Consequently, it is expected that such CN deployment seriously affects the improvement of the assumed dynamic radio resource allocation scheme.

Table 11.1 Experiment conditions.

Frequency	4849.98 MHz
Bandwidth	99.72 MHz
RU specifications	Downlink: 4 × 4 MIMO/4-diversity Uplink: 1 × 1 MIMO/4-diversity
Modulation scheme	QPSK/16QAM/64QAM/256QAM; Fixed to QPSK in the experiment

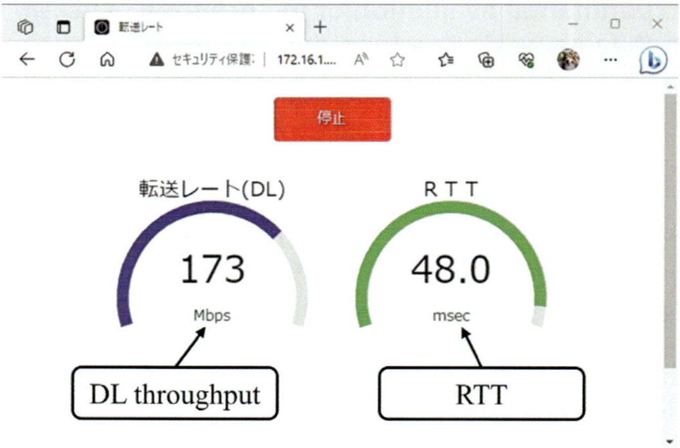

(a)

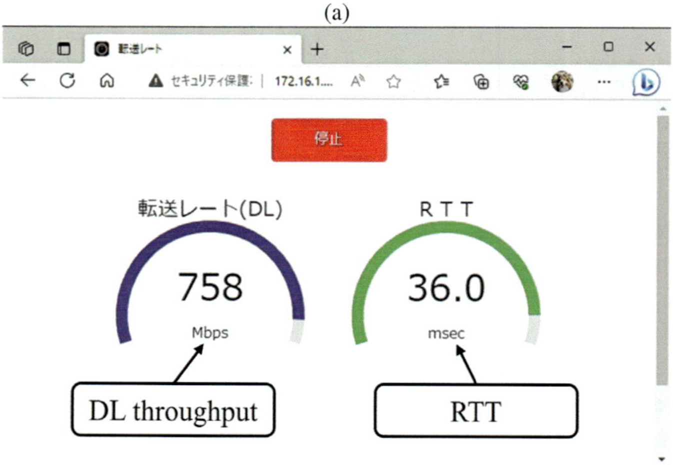

(b)

Figure 11.9 Examples of the performances observed on the developed applications. (a) DL performance employing the core in Hokuriku-site. (b) DL performance employing the core in Tokyo-site.

Table 11.2 Results of the experiments.

Performances for CN located in Hokuriku-site	
Median of DL throughput	250 Mb/s
Median of RTT	60 ms
Performances for CN located in Tokyo-site	
Median of DL throughput	800 Mb/s
Median of RTT	25 ms

11.5 Expectations on the Further Advanced System Evaluations

11.5.1 Concept of multiple base station environments employing the slow adaptive modulation technology

In this section, considerations on the further advanced utilization of the mobile service areas in B5G applications are described, which is also assumed to be evaluated in the B5G mobile environment of B5G/IoT testbed with high reliability and high elasticity.

As an example of the possible enhancement for such mobile area utilization in B5G systems, the deployment of multiple BSs environments by employing the slow adaptive modulation technology [10] is considered in this chapter. Figure 11.10 shows a concept of the slow adaptive

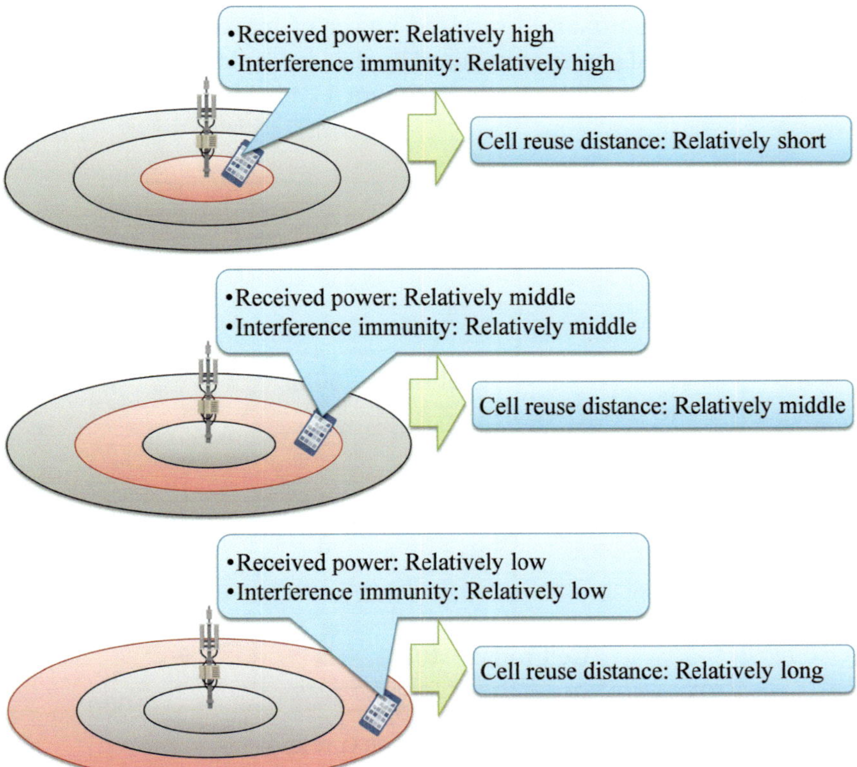

Figure 11.10 A concept of the slow adaptive modulation scheme with dynamic channel assignment.

modulation scheme with dynamic channel assignment. This technology effectively exploits the radio communication conditions owing to the distance from the base station thereby suitable control of the cell reuse distance among base stations, as depicted in Figure 11.10.

11.5.2 Numerical evaluation of the proposed system on the mobile network development evaluation

In order to investigate the basic performance of such multiple BSs deployed situation, numerical evaluation on the normalized radio resource performances according to the assumed cell reuse distance as shown in Figure 11.11. Normalized radio resource is defined as the allocated resource per cell as in Figure 11.10, which is normalized by the total resource amount of the system. In the figure, the normalized radio resource characteristics according to the various cell radii and cell reuse distance are provided. Characteristics for normalized received power of the signal that is defined as the received signal power normalized by that for deployment cases of cells with a 40-m radius are also shown in Figure 11.11. It is confirmed that the

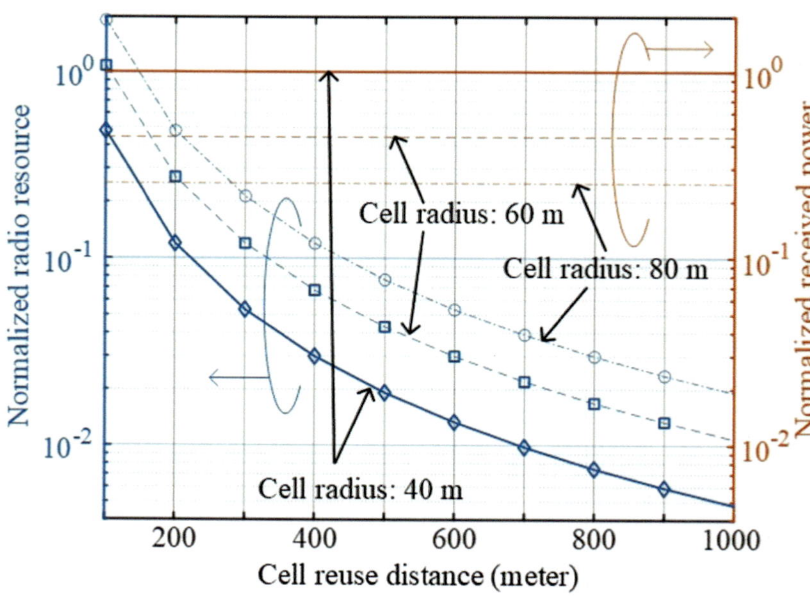

Figure 11.11 Normalized radio resource performances for the assumed system evaluations in the B5G mobile evaluation environment.

effective utilization of the radio resources is affected by both the assumed cell radii and the cell reuse distance as shown in the radio resource performance in Figure 11.11, and the configurations of such system parameters become very important, especially in the promotion of B5G systems and services in the future.

To further evaluate the assumed system performances by employing the slow adaptive modulation scheme as depicted in Figure 11.10, parameters of the slow adaptive modulation scheme are introduced in this subsection. The evaluation assumes higher gain modulation allocation in the central region of each cell. The central region is defined by a parameter R_{sams} that reveals a normalized radius of the cell ($0 < = R_{sams} < = 1$). Furthermore, the assumed higher gain modulation is also realized by another parameter G_{sams} that reveals a carried traffic expected on the assumed radio resource in the cell central region, normalized by that for the outer region. For example, $G_{sams} = 2.0$ employed case means the situation where two times as large traffic is carried by the assumed radio resource in the cell center region compared with the outer region. Figure 11.12 summarizes the parameters employed in the performance evaluations.

Figures 11.13 and 11.14 show the evaluation results. In the evaluations, uniform distributions of the mobile terminals in each cell are assumed. Figures 11.13 and 11.14 show the averaged performances of the normalized

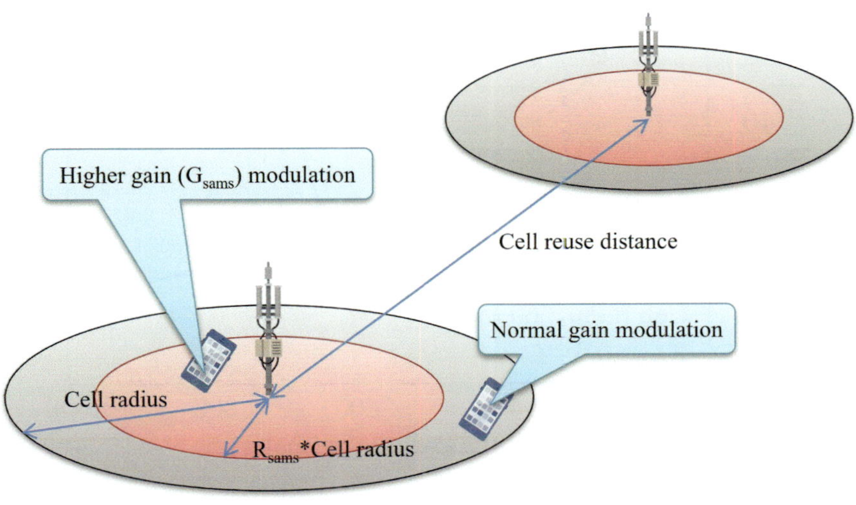

Figure 11.12 Parameters employed in the evaluations of the slow adaptive modulation scheme.

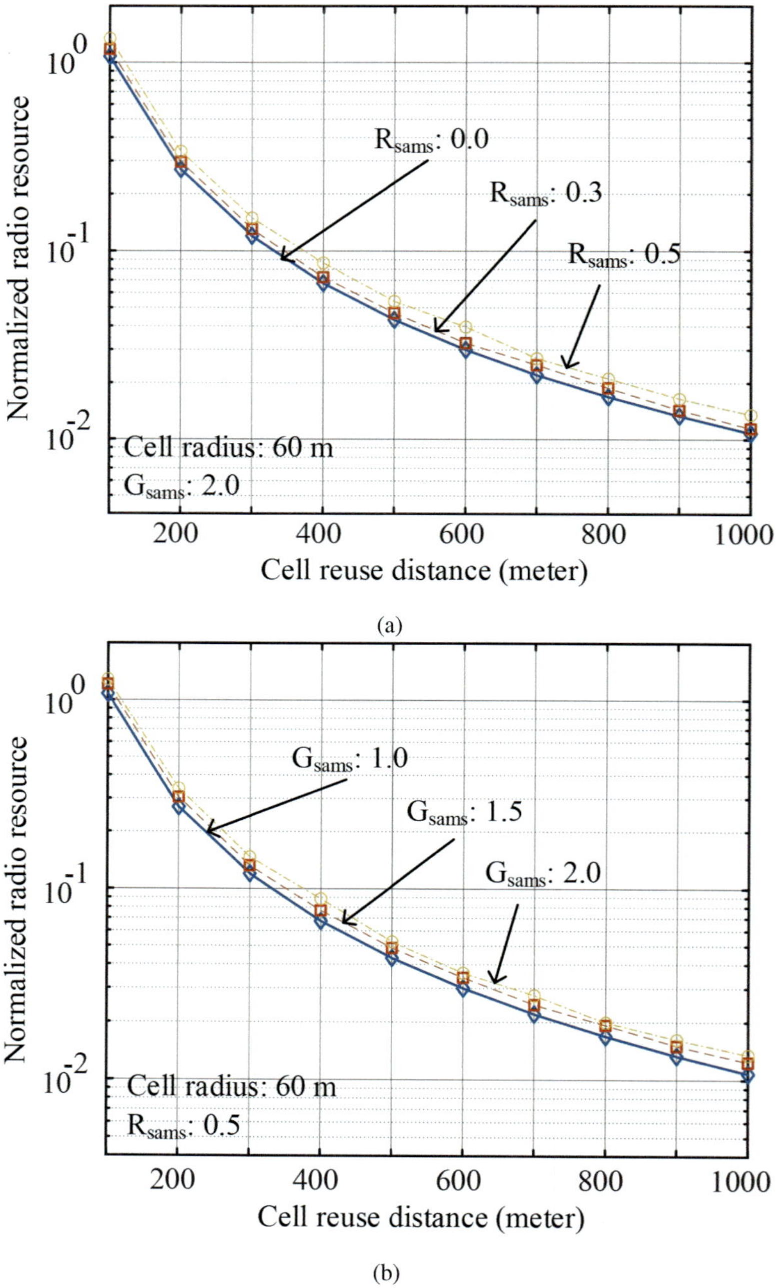

Figure 11.13 Performance evaluations according to the parameters of slow adaptive modulation scheme. (a) R_{sams} as a parameter. (b) G_{sams} as a parameter.

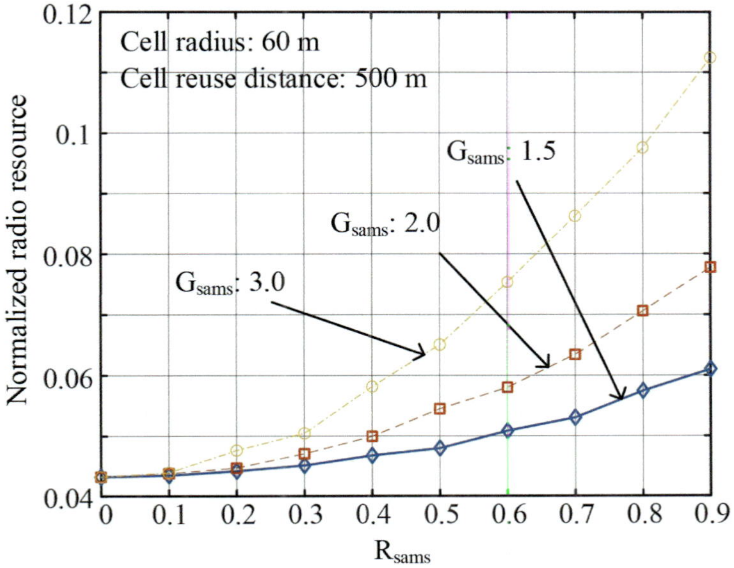

Figure 11.14 Performance evaluations according to both parameters of R_{sams} and G_{sams}.

radio resource obtained by all mobile terminals. Those figures confirm that the higher gain modulation employed in the central area provides the higher normalized radio resource performance for the entire system, as in 120% improvement in the normalized radio resource when the slow adaptive modulation technology is employed with the normalized cell radius parameter $R_{sams} = 0.5$ and the normalized gain parameter $G_{sams} = 2.0$, compared to the case without the slow adaptive modulation technology ($R_{sams} = 0.0$, $G_{sams} = 1.0$), as confirmed in Figure 11.13 situation. However, as shown in Figure 11.10, such higher gain modulation that requires higher received power and higher interference immunity conditions surely restricts the available region. In order to configure suitable system performances, optimizations of those system parameters are essential issues. Developed functions in the B5G mobile environment are expected to be effectively exploited in such future mobile system evaluations.

11.6 Conclusion

In this chapter, a dynamic radio resource allocation scheme according to the radio utility conditions for B5G mobile applications is proposed and evaluated in the B5G/IoT testbed environment with high reliability and high elasticity. The basic performances of the proposed scheme are evaluated on the testbed environment that confirms the effectiveness of the proposed dynamic resource allocation. The chapter also considers the further advanced B5G mobile application under multiple BS deployment situations. Investigations via those experiments and computer simulations confirm that such diversified wireless access capabilities and software-based customization functions provided as reliability and elasticity performances in the testbed effectively work to evaluate the highly advanced and complicated systems and services in B5G society. Further detailed considerations on the effective radio resource utilization scheme accompanied by the evaluation of the testbed environment are considered as the future study.

Acknowledgments

This research has been conducted under the contracts "R&D for the realization of high-precision radio wave emulator in cyberspace" and "R&D on the deployment of mobile communication systems in the millimeter wave band and other frequencies" (JPJ000254) made with the Ministry of Internal Affairs and Communications of Japan.

References

[1] ITU-R, "IMT Vision - Framework and overall objectives of the future development of IMT for 2020 and beyond," ITU-R Recommendations M.2083-0, 2015.

[2] DARPA, "World's Most Powerful RF Emulator to Become National Wireless Research Asset: The Colosseum - a massive RF channel emulator used to run DARPA's Spectrum Collaboration Challenge - transitions to National Science Foundation's PAWR Program after October finale," https://www.darpa.mil/news-events/2019-09-03

[3] DARPA, "Spectrum Collaboration Challenge (SC2)," https://archive.darpa.mil/sc2/

[4] Takeshi Matsumura et al., "Platform for wireless system emulation based on cyber physical fusion," IEICE Technical Report, SRW2021-26, 2021.

[5] NICT, "JGN Website High Speed R&D Network Testbed," https://testbed.nict.go.jp/jgn/english/index.html

[6] NICT, "StarBED5 Project website," https://starbed.nict.go.jp/en/
[7] Hiroaki Harai, "Beyond 100 Gbps R&D Network Testbed with Optical Technology," Proceedings of the OECC 2019, pp. 1-3, 2019.
[8] Hiroaki Yamanaka et al., "Implementation and Evaluation of an Edge Computing Testbed Simplifying Creation of Experimental Environments," IEICE Technical Report, CQ2020-91, 2020
[9] Hiroaki Harai et al., "Deployment of Circular Evolutionary Testbed toward Beyond 5G Mobile Network," IEICE Technical Report, NS2022-63, 2022.
[10] T. Ikeda, S. Sampei, and N. Morinaga, "TDMA-based adaptive modulation with dynamic channel assignment for high-capacity communication systems," IEEE Trans. Vehicular Technology, vol. 49, no. 2, pp. 404-412, March 2000.
[11] S. Sampei, S. Komaki, and N. Morinaga, "Adaptive modulation/TDMA scheme for large capacity personal multi-media communication systems", IEICE Trans. Commun., vol. E77-B, no. 9, pp. 1096-1103, Sept. 1994.
[12] F. Kojima, T. Miyachi, T. Matsumura, H. Sawada, H. Harai and H. Harada, "A Large-Scale Wireless Emulation Environment with Interaction between Physical and Virtual Radio Nodes for Beyond 5G Systems," 2022 IEEE 33rd Annual International Symposium on Personal, Indoor and Mobile Radio Communications (PIMRC), Kyoto, Japan, 2022, pp. 1-6
[13] T. Matsumura, K. Ibuka, H. Murakami, K. Ishizu, and F. Kojima, "A Prototype of Multi-RAN eNB with Dynamic Baseband Resource Allocation for Heterogeneous Private LTE Networks," 21st International Symposium on Wireless Personal Multimedia Communications (WPMC2018), pp. 46-51, Nov. 2018.
[14] T. Miyachi et al., "CyReal: A Unified Framework for Continuous Evaluation of Simulations, Emulations, and Real Systems," 16th EAI International Conference on Simulation Tools and Techniques (EAI SIMUtools 2024), Dec. 2024.

Part II

Applications

DOI: 10.1201/9788743801108-13

12

Quantum Machine Learning – The Need of Time

Vandana Rohokale[1] and Ramjee Prasad[2]

[1]Sinhgad Institute of Technology and Science, India
[2]CTIF Global Capsule, Herning, Denmark
E-mail: vmr.301075@gmail.com; prasad.ramjee@ctifglobalcapsule.org

Abstract

Quantum machine learning (QML) represents the fusion of quantum computing algorithms with machine learning techniques. A prominent application of QML involves employing machine learning algorithms to analyze classical data using a quantum computer, known as quantum-enhanced machine learning. This interdisciplinary field merges concepts from quantum physics and machine learning to pioneer new algorithms and methodologies for data processing and analysis. QML seeks to harness unique quantum properties like superposition and entanglement to enhance the performance of machine learning models. It explores the synergy between quantum computing and machine learning, aiming to leverage quantum systems' inherent characteristics for advanced data processing capabilities. This chapter provides an in-depth exploration of QML's comprehensive approach.

Keywords: Quantum machine learning (QML), superposition and entanglement, machine learning, quantum computing

12.1 Introduction

In the modern era, traditional computational techniques often face insurmountable challenges due to technical limitations and computational complexity. Quantum computing emerges as a pivotal solution in such

DOI: 10.1201/9788743801108-14

scenarios. Beyond these hurdles, domains like machine learning and biochemistry hold promise for leveraging existing quantum devices. Newly developed algorithms often represent quantum adaptations of familiar techniques such as support vector machines and deep learning methods like quantum neural networks.

Currently, numerous researchers are dedicated to overcoming traditional machine-learning challenges using quantum devices and methodologies. The field of quantum machine learning (QML) is actively advancing, though it requires time to fully realize its potential benefits. Despite ongoing progress, today's quantum computers still encounter issues related to quality, speed, and scalability. Therefore, intensive research into quantum hardware and software co-design remains essential for future advancements [1].

Quantum machine learning (QML) holds profound implications for both quantum computing and artificial intelligence (AI), promising to transform the way we handle and analyze intricate data sets. This potential breakthrough could lead to significant advancements in tackling complex problems. For instance, in the realm of classification problems, the exponential runtime and suboptimal results often associated with traditional approaches can be mitigated by leveraging quantum computing techniques alongside machine learning algorithms. Quantum circuits offer a novel approach to encoding data, while quantum kernels show promising potential to drastically accelerate certain types of classification tasks compared to conventional methods. Here are several key points underscoring the importance and potential of QML [2, 3].

1. Enhanced computational power: quantum computing provides exponential computational power compared to classical methods, enabling the exploration of vast solution spaces and more efficient algorithms. QML utilizes this capability to address complex AI tasks and optimize machine learning models.
2. Enhanced autonomy through quantum properties: QML harnesses the unique properties of quantum systems, such as superposition and entanglement, to enhance machine learning models. This enables simultaneous exploration of multiple solutions, potentially leading to faster and more accurate computations.
3. Improved optimization capabilities: many AI challenges involve optimization tasks, such as parameter tuning and system optimization. Quantum algorithms and optimization techniques leverage quantum principles to enhance efficiency in exploring solution spaces.

4. Advanced data analysis: quantum data exhibits quantum behavior and often requires specialized analysis techniques. QML algorithms enable efficient processing and analysis of quantum data, advancing our understanding of quantum systems in fields like physics, chemistry, and material science.
5. Capability to solve complex problems: QML has the potential to tackle problems that are challenging for traditional machine learning algorithms or computationally impractical. This includes large-scale optimizations, complex simulations, and analysis of quantum data to reveal patterns and insights that were previously difficult to extract.
6. Real-world applications: QML introduces new avenues for solving real-world problems across various domains. Applications include simulating molecular interactions for drug discovery, optimizing financial portfolios, and improving quantum communication technologies.

12.2 Fundamentals of Quantum Computing

The fundamentals of quantum computing involve understanding the basic principles and concepts that underpin the field. Here are the key fundamentals of quantum computing [4, 5].

1. *Quantum bits (qubits)*

The fundamental building block of quantum computing is the quantum bit or qubit. Contrasting to classical bits, representing either 0 or 1, qubits exist in a superposition of both states simultaneously. Because of this property, qubits can perform parallel computations and enable quantum computers to process huge amounts of information in parallel. Here are the key characteristics of the quantum bits.

Superposition: qubits can be in a state of superposition, meaning they can represent a combination of 0 and 1 simultaneously. This superposition is described using complex numbers and can be represented mathematically as a linear combination of the two basis states, $|0\rangle$ and $|1\rangle$. For example, a qubit can be in a state $\alpha|0\rangle + \beta|1\rangle$, where α and β are complex numbers representing the probability amplitudes.

Coherent manipulation: qubits can be exploited through quantum gates that are similar to traditional logic gates. Quantum gates allow for the transformation of qubit states and the performance of quantum computations. Universal quantum gates are the Hadamard gate, Pauli gates (X, Y, Z), CNOT gate, etc.

Measurement: while measuring a qubit, it collapses into one of the two traditional states (0 or 1) with the probability decided by the superposition coefficients. The measurement process provides a classical outcome, extracting information from the quantum system. The probabilities of measurement outcomes are related to the squared magnitudes of the probability amplitudes.

Entanglement: qubits can be entangled, forming associations between their states even when physically separated. Entanglement is a unique property of quantum mechanics that enables the representation of complex relationships between qubits. It allows for highly efficient information processing and communication in quantum systems and plays a crucial role in quantum algorithms and quantum communication protocols.

Quantum parallelism: superposition allows qubits to exist in multiple states simultaneously, enabling quantum parallelism. Quantum algorithms can exploit this parallelism to perform computations on all possible inputs simultaneously, potentially leading to exponential speedup compared to classical algorithms.

Feebleness and decoherence: qubits are fragile and prone to environmental noise and decoherence. Interactions with the surrounding environment can cause the loss of quantum information, leading to errors in computations. Overcoming these challenges is a crucial aspect of quantum computing research.

The properties and behavior of qubits enable quantum computers to perform computations and solve certain problems more efficiently than classical computers. By harnessing the principles of superposition, entanglement, and quantum interference, qubits form the foundation for the power and potential of quantum computing.

2. *Quantum gates*

Quantum gates are basic building blocks in quantum computing that allow for the exploitation and transformation of quantum bits (qubits). Like conventional logic gates, quantum gates are used to perform operations on qubits, enabling various computational tasks [6, 7, 8]. However, quantum gates have distinct properties due to the principles of quantum mechanics. Figure 12.1 shows various types of quantum gates.

Here are some key characteristics of the quantum gates.

Unitary transformations: quantum gates are represented with unitary transformations, which are reversible operations that preserve the normalization of the quantum state. Unitary transformations are shown by complex matrices,

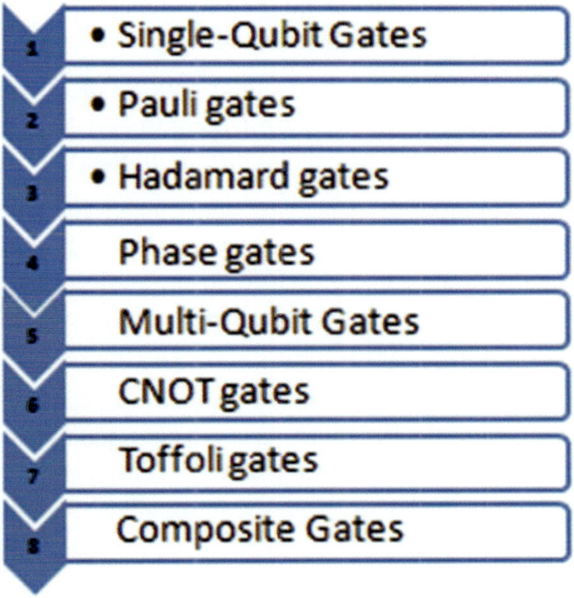

Figure 12.1 Quantum gates.

and applying a unitary gate to a qubit corresponds to multiplying the quantum state vector by the corresponding unitary matrix.

Single-qubit gates: single-qubit gates operate on individual qubits. They can manipulate the state of a single qubit without affecting other qubits in a quantum system. Common examples of single-qubit gates include

Pauli gates: the Pauli gates include the X gate (bit-flip), Y gate (bit- and phase-flip), and Z gate (phase-flip). These gates rotate the qubit state around the X, Y, or Z axes of the Bloch sphere.

Hadamard gate: the Hadamard gate (H gate) creates superposition by transforming the basis states $|0\rangle$ and $|1\rangle$ into an equal superposition state.

Phase gate: the phase gate (S gate) introduces a phase shift of $\pi/2$ to the $|1\rangle$ state.

Multiqubit gates: multiqubit gates are used for carrying out operations on multiple qubits simultaneously, allowing for the generation of entanglement and enabling more complex computations. Common multiqubit gates include

CNOT gate: the controlled-NOT (CNOT) gate is a two-qubit gate that turns over the second qubit (the target qubit) if the first qubit (the control qubit)

is in the $|1\rangle$ state. It is a fundamental gate for generating entanglement and implementing various quantum algorithms.

Toffoli gate: the Toffoli gate is a three-qubit gate which is called a universal reversible logic gate that acts as a controlled-controlled-NOT gate. It flips the third qubit if both the first and second qubits are in the $|1\rangle$ state.

Composite gates: composite gates are constructed by combining multiple single-qubit gates and/or multiqubit gates. By combining these gates, more complex operations can be performed on qubits. Common examples of composite gates include the controlled-phase gate (CPhase) and the controlled-rotation gate (CROT).

Quantum gate decomposition: arbitrary unitary operations can be decomposed into a sequence of elementary gates. This decomposition allows for the construction of any unitary transformation using a combination of basic gates. Gate decomposition is crucial for optimizing quantum circuits and implementing quantum algorithms efficiently.

No-cloning theorem: one important aspect of quantum gates is that they cannot perfectly clone an unknown quantum state. This is known as the no-cloning theorem and is a fundamental principle in quantum information theory.

Quantum gates play a critical role in quantum algorithms and quantum circuits. By combining different gates and manipulating the state of qubits, complex computations and quantum information processing tasks can be performed.

12.3 Comparison of Quantum Computing and Classical Computing

Quantum computing and Classical computing are two fundamentally different exemplars of computation [9, 10, 11]. Here is a comparison of key aspects between quantum computing and classical computing as shown in Table 12.1.

12.4 Machine Learning

Classical machine learning algorithms are computational methods designed to analyze data and make predictions or decisions [12, 13]. These algorithms are tailored for classical computers and have been widely utilized across various domains. Here are some commonly employed classical machine learning algorithms.

Table 12.1 The comparison of classical and quantum computing.

Sr. no./details	Classical computing	Quantum computing
Basic units of information	It uses bits as the basic units of information. Bits can be represented either as 0 or 1.	It uses qubits as the basic units of information. Qubits can exist in a superposition of 0 and 1, enabling parallel computations.
Representation and processing	It processes the information using classical logic gates, which manipulate bits based on predetermined rules.	It processes information using quantum gates, which operate on qubits and make use of quantum techniques such as superposition and entanglement.
Information storage	It stores and manipulates information in binary form, using physical elements such as transistors, magnetic storage devices, or optical media.	It stores and manipulates information in the quantum state of qubits. The qubit's state is typically maintained using quantum systems such as trapped ions, superconducting circuits, or topological properties.
Computation power	It performs computations sequentially, following a set of instructions. The computational power of classical computers increases with the number of bits in the computation.	It performs calculations considerably faster than classical computers for specific problems. It supports parallelism and interference effects to explore manifold computational paths simultaneously, potentially providing significant speedup for many algorithms.
Error and noise	These are deterministic and have error correction mechanisms that can effectively detect and correct errors.	These are susceptible to errors and noise due to the fragile nature of qubits. Error correction techniques in QML are developed to diminish errors and protect quantum information.
Applications	These are extensively used in various domains, including data processing, simulations, cryptography, artificial intelligence, and everyday computing tasks.	QML applications have the potential to impact fields such as cryptography, optimization, quantum chemistry, material science, and machine learning.

Linear regression: a supervised learning algorithm used for regression tasks, linear regression prototypes the relationship between a dependent variable and one or more independent variables by fitting a linear equation to the data.

Logistic regression: also a supervised learning algorithm, logistic regression is utilized for classification tasks. It models the relationship between a dependent variable and independent variables by fitting a logistic function to the data.

Decision trees: these famous supervised learning algorithms can handle both classification and regression tasks. They construct a tree-like model of decisions and their potential outcomes based on data features.

Random forest: an ensemble learning algorithm that combines multiple decision trees, reducing overfitting and enhancing generalization by aggregating predictions from individual trees.

Support vector machines (SVM): a powerful supervised learning algorithm for both classification and regression tasks, SVM identifies the optimal hyperplane that separates data points of different classes with the maximum margin.

Naive Bayes: a probabilistic supervised learning algorithm primarily used for classification tasks. Naive Bayes relies on Bayes' theorem and assumes independence between features, making it efficient for large data sets.

K-nearest neighbors (k-NN): an effective unsupervised learning algorithm for both classification and regression tasks, k-NN predicts the value of a data point based on the majority vote or average of its k nearest neighbors.

Neural networks: inspired by the human brain's structure and function, neural networks consist of interconnected nodes (neurons) organized in layers. They are versatile and can be applied to various tasks such as classification, regression, and deep learning.

Gradient boosting algorithms: examples include gradient boosting machines (GBM), XGBoost, and LightGBM, which are ensemble learning methods. They sequentially combine weak learners (typically decision trees), iteratively improving predictive performance by correcting the errors of previous models.

These algorithms collectively form the backbone of classical machine learning, empowering applications across diverse fields through their ability to analyze and derive insights from data.

12.4.1 Challenges and limitations

Classical machine learning has demonstrated its utility across various domains but is also confronted with several challenges and constraints [14]. Here are some of the primary challenges and limitations encountered in classical machine learning:

Dimensionality: traditional machine learning algorithms often struggle with high-dimensional data. As the number of features increases, the amount of required data grows exponentially, leading to heightened computational complexity, overfitting, and difficulties in identifying meaningful patterns.

Feature engineering: classical machine learning heavily relies on manual feature engineering, where domain experts manually extract relevant features from raw data. This process can be time-consuming, subjective, and may limit the algorithm's ability to autonomously discover complex and abstract features.

Generalization issues: classical machine learning algorithms typically generalize well to unseen data if the training data represents the underlying distribution accurately. However, they often struggle with out-of-distribution data that significantly differs from the training set, resulting in reduced performance in real-world scenarios where data distributions evolve over time.

Limited learning capacity: traditional machine learning algorithms have a finite learning capacity, making them less effective for complex tasks that involve capturing intricate relationships in data. They may underperform when faced with highly nonlinear, hierarchical, or context-dependent patterns.

Data sensitivity: the performance of classical machine learning algorithms heavily depends on the quality and quantity of training data. Insufficient or noisy data can lead to poor model performance and unreliable predictions, while biased data sets can skew the learning process and outcomes.

Interpretability and explainability: many classical machine learning algorithms, such as deep neural networks, are considered black-box models, making it challenging to interpret and explain their decisions. This lack of transparency can hinder trust, accountability, and acceptance of models, particularly in sensitive domains like healthcare or finance.

Overfitting and underfitting: overfitting occurs when a model captures noise and irrelevant patterns from the training data, while underfitting happens when a model is too simplistic to capture the complexities of the data. Balancing between these extremes is crucial in classical machine learning.

Scalability: traditional machine learning algorithms may encounter scalability issues when dealing with large-scale data sets or high-dimensional feature spaces. Training and inference times can become impractical, limiting their application to big data problems.

Contextual understanding: classical machine learning algorithms primarily focus on pattern recognition without deep contextual understanding. They may struggle with tasks requiring common-sense reasoning, causal inference, or nuanced contextual comprehension, leading to suboptimal performance in such scenarios.

Adaptability: traditional machine learning models are static and require retraining from scratch to adapt to new data or evolving environments. They lack real-time adaptability, which can be a limitation in dynamic systems or rapidly changing data scenarios.

These challenges underscore the ongoing need for advancements in machine learning techniques to overcome these limitations and enhance the applicability of AI across diverse and complex real-world problems.

12.5 Bridging Quantum Computing and Machine Learning

Quantum-enhanced techniques for feature selection and dimensionality reduction leverage quantum computing to enhance the efficiency and effectiveness of these processes in traditional machine-learning tasks [13, 14]. Here is an overview of how quantum computing can improve feature selection and dimensionality reduction.

Feature selection: feature selection involves identifying a subset of relevant features from a larger set, eliminating irrelevant or redundant information. Quantum computing offers potential advantages in this area by employing quantum algorithms to explore the combinatorial space of feature subsets more efficiently.

Quantum feature selection (QFS): QFS algorithms harness quantum parallelism and quantum search algorithms like Grover's algorithm to accelerate the exploration of feature combinations. Leveraging quantum superposition, these algorithms aim to identify relevant features with fewer evaluations compared to classical methods.

Dimensionality reduction: dimensionality reduction techniques aim to transform high-dimensional data into a lower-dimensional representation while preserving essential patterns and relationships. Quantum computing has the potential to enhance dimensionality reduction by utilizing quantum algorithms to perform more efficient computations on high-dimensional data.

Quantum principal component analysis (QPCA): QPCA is a quantum adaptation of the classical principal component analysis (PCA) algorithm. It utilizes quantum algorithms to compute principal components from high-dimensional data, facilitating faster dimensionality reduction. QPCA may offer speedups over classical PCA for large and complex data sets.

Quantum singular value decomposition (QSVD): QSVD is a quantum algorithm designed to compute the singular value decomposition of matrices, crucial for methods like truncated SVD in dimensionality reduction. By leveraging quantum computations, QSVD aims to accelerate the dimensionality reduction process and enhance data representation.

Using Qiskit runtime, developers can easily construct quantum machine learning algorithms with tools like the sampler primitive, unique to IBM's Qiskit runtime. These primitives are predefined programs that optimize workflows and execute efficiently on quantum systems.

12.6 QML in Cyber Security

Technical failures and rare catastrophic events often pose significant threats to the confidentiality, integrity, and availability of computing systems. Cyberattacks manifest in various threat scenarios including espionage, organized crime, hacktivism, and cyber warfare. These attacks typically adhere to common patterns illustrated in Figure 12.2, depicting distinct phases of cyber operations. The following text elaborates on these phases.

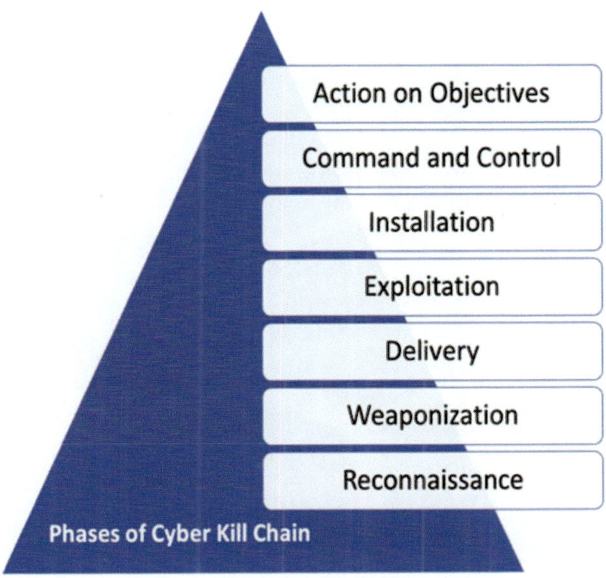

Figure 12.2 Phases of sophisticated cyberattacks in the cyber kill chain.

- *Reconnaissance phase*: this phase involves passive or active research conducted on the target.
- *Weaponization phase*: here, tools are developed or adapted for use in attacks.
- *Delivery phase*: the weaponized tool or payload is delivered to the target.
- *Exploitation phase*: access to the victim system is gained, and malware is prepared for deployment.
- *Installation phase*: malware is successfully installed on the victim system.
- *Command and control phase*: remote control over the victim's system is established.
- *Action on objectives phase*: This phase focuses on achieving the goals of the cyberattack.

Cyber security is crucial for ensuring the confidentiality, integrity, and availability of IT data, protecting it from various threats. With the increasing volume and complexity of data, training deep models has become more costly and time-consuming. Quantum machine learning (QML) offers potential benefits such as faster processing, reduced energy consumption, and cost-effective algorithms for detecting cyberattacks and implementing robust security solutions. Figure 12.3 illustrates various potential applications of QML in enhancing cyber security.

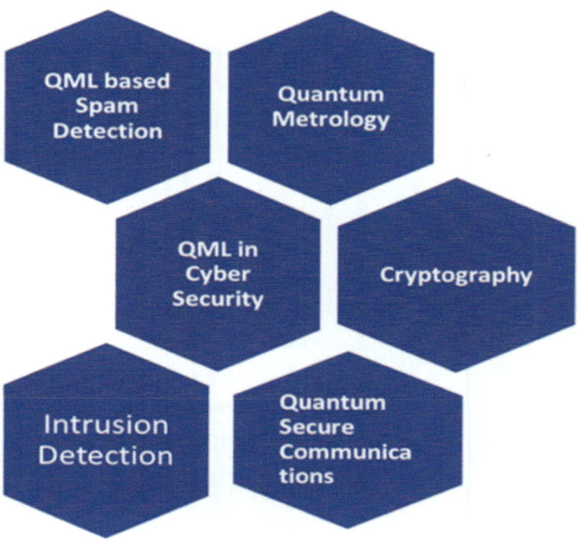

Figure 12.3 QML in cyber security.

1. *QML in spam detection*

Spam refers to unsolicited commercial electronic messages, such as advertisements, business proposals, or financial offers, often used maliciously for purposes like phishing. This involves impersonation to deceive users into divulging sensitive information like banking credentials. Spam can originate from various sources including social networks, product reviews, blogs, and microblogs. An emerging form involves fake reviews. Machine learning techniques have been extensively researched by service providers like Yahoo, Google, and Microsoft to filter spam emails. Methods include probabilistic approaches, decision trees, artificial immune systems, support vector machines, and artificial neural networks. Deep learning and adversarial learning are also employed to combat spam threats effectively.

Quantum machine learning (QML) is increasingly recognized as pivotal in cybersecurity. Key applications include spam detection, malware detection, and security based on user or network behavior. Both attackers and defenders view QML as essential for strategy formulation. Spam detection, a longstanding issue, is now leveraging machine learning algorithms, with ongoing exploration into QML-based tools such as text classifiers.

Quantum metrology utilizes precise particle manipulation for high-accuracy measurements, facilitating advancements in radar and camera technology. These innovations are critical in defense and security applications, enhancing capabilities to detect clandestine aircraft or unmanned aerial vehicles. QML could potentially develop new location detection mechanisms independent of GPS signals, which are susceptible to interference.

Cryptography, fundamental to security, relies on random number generation. Quantum technology surpasses traditional methods by leveraging quantum physics to generate truly random numbers. This unpredictability ensures robust security, essential in cryptography applications.

Quantum secure communications utilize quantum principles for safeguarding critical infrastructure. Quantum key distribution, for instance, employs reduced laser pulses to share encryption keys securely between sender and receiver. This technology is crucial for securing sensitive government and mission-critical communications, laying the groundwork for future quantum computer networks.

12.7 Benefits and Applications of QML

Quantum computers are not meant to replace traditional classical computers but rather excel at specific computationally intensive tasks that classical

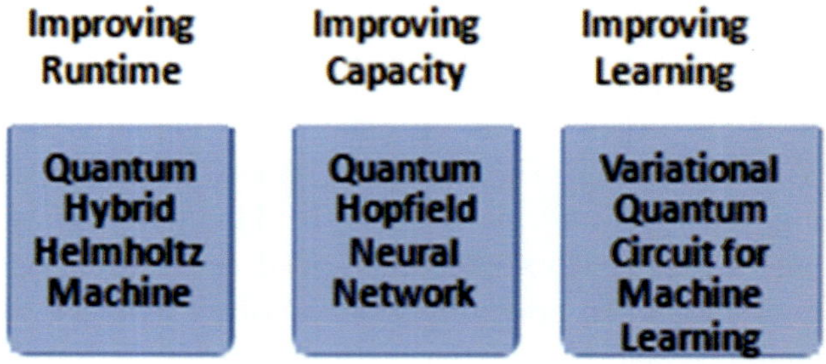

Figure 12.4 Benefits of quantum machine learning.

computers find challenging. Hybrid computing, where quantum computers handle specific tasks while traditional computers manage routine operations, is gaining popularity. Figure 12.4 illustrates the advantages of quantum machine learning (QML), highlighting enhancements in runtime, capacity, and learning efficiency achieved through various QML techniques such as the quantum hybrid Helmholtz machine, quantum Hopfield neural network, and variational quantum circuit for machine learning [15].

Figure 12.5 outlines diverse applications of quantum machine learning, including deep study of nanoparticles, development of new materials through molecular and atomic mapping, molecular modeling for drug discovery and medical research, exploration of the human body's intricate structure, improvements in pattern recognition and classification, advanced space exploration, applications in banking and finance, and comprehensive cybersecurity solutions.

- Drug discovery and materials science: quantum machine learning (QML) accelerates drug discovery and materials science research by simulating molecular behavior, predicting properties, and designing new compounds. Quantum algorithms like variational quantum eigensolvers (VQE) optimize molecular structures and enhance quantum chemistry calculations efficiently.
- Financial modeling and portfolio optimization: QML algorithms support financial modeling and portfolio optimization tasks. Quantum algorithms such as quantum support vector machines (QSVM) and quantum neural networks (QNN) improve predictions and risk assessments in financial markets. They also optimize investment portfolios and asset allocation strategies.
- Image and speech recognition: QML enhances image and speech recognition by processing large data sets with quantum algorithms. This improves accuracy in tasks like classification, pattern recognition, and

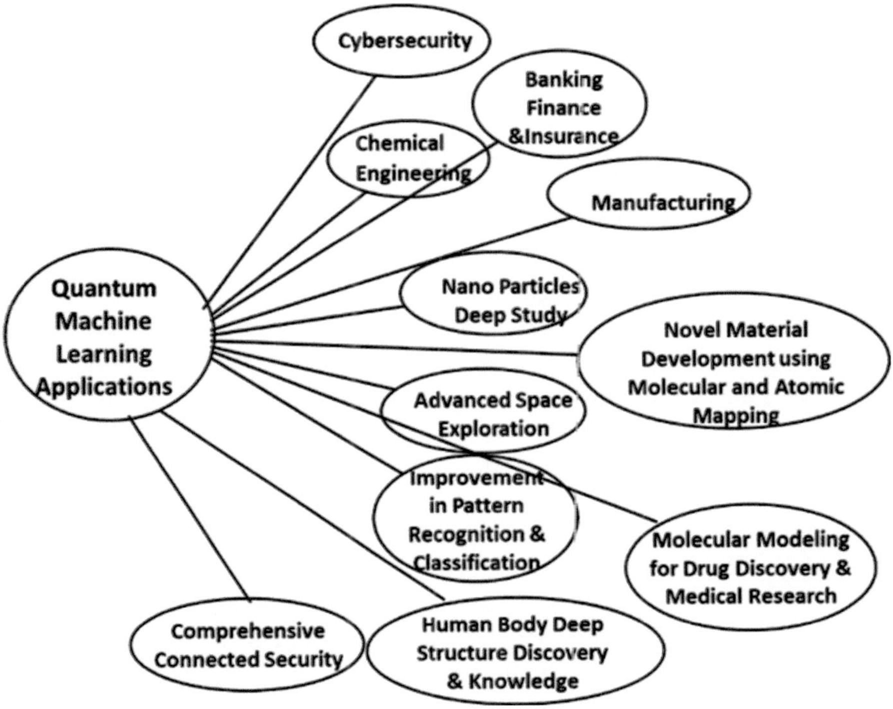

Figure 12.5 Applications of QML.

feature extraction, benefiting applications in computer vision, natural language processing, and speech synthesis.

- Optimization and operations research: quantum algorithms such as quantum approximate optimization algorithm (QAOA) and quantum annealing solve complex optimization problems. They find applications in supply chain management, logistics, scheduling, and resource allocation where optimal solutions are critical.
- Quantum-enhanced machine learning: QML algorithms enhance traditional machine learning techniques by using quantum resources for feature selection, dimensionality reduction, and model training. Quantum feature selection and dimensionality reduction techniques improve data representation and reduce computational complexity in machine learning tasks.
- Quantum key distribution: quantum key distribution leverages quantum entanglement for secure cryptography, where changes to one entangled

particle instantly affect its partner regardless of distance, ensuring secure communication. Companies like KetQuantum and OpenQuantum are pioneers in quantum cryptography.

12.8 Challenges and Future Scope

Here are some key challenges and future prospects for quantum machine learning (QML).

Hardware limitations: a significant challenge is the current state of quantum hardware. Existing quantum computers suffer from limited qubit coherence, high error rates, and restricted qubit connectivity. Overcoming these limitations and scaling up qubit numbers while enhancing their reliability is essential to fully unlock the potential of QML.

Noise and error correction: quantum systems are prone to noise and errors. Developing effective error correction techniques and strategies to mitigate errors is critical to improving the accuracy and reliability of QML algorithms. Research in quantum error correction codes and fault-tolerant quantum computing is actively addressing these challenges.

Algorithm development and optimization: designing efficient quantum algorithms tailored for specific machine learning tasks remains challenging. Ongoing research focuses on developing new quantum algorithms, refining existing ones, and optimizing their performance. This includes exploring quantum circuits, quantum-inspired classical algorithms, and hybrid classical-quantum approaches to enhance machine learning capabilities.

Quantum data representation and preprocessing: efficient methods for encoding and processing classical data into quantum form are actively being researched. Developing effective quantum data representation mechanisms is crucial to leveraging quantum algorithms and harnessing the unique properties of quantum systems.

Validation and benchmarking: validating and benchmarking QML algorithms against classical counterparts and real-world data sets is essential. Establishing metrics and testbeds to evaluate the accuracy, scalability, and applicability of QML algorithms is crucial for their practical adoption and effectiveness in various applications.

References

[1] R. Divya and J. Dinesh Peter, "Quantum Machine Learning: An comprehensive review on optimization of machine learning algorithms," 2021

Fourth International Conference on Microelectronics, Signals & Systems (ICMSS), Kollam, India, 2021, pp. 1-6, doi:10.1109/ICMSS53060.2021.9673630.

[2] S. N. Pushpak and S. Jain, "An Introduction To Quantum Machine Learning Techniques," 2021 9th International Conference on Reliability, Infocom Technologies and Optimization (Trends and Future Directions) (ICRITO), Noida, India, 2021, pp. 1-6, doi:10.1109/ICRITO51393.2021.9596240.

[3] K. Bharti, V. Vedral, T. Haug and L.-C. Kwek, "Machine Learning meets Quantum Foundations: A Brief Survey", AVS Quantum Sci. 2 034101, vol. 2, no. 2639–0213, pp. 17, 2020.

[4] E. Aimeur, G. Brassard and S. Gambs, "Machine learning in a quantum world" in Advances in Artificial Intelligence, Springer, pp. 431-442, 2006.

[5] M. Schuld, I. Sinayskiy and F. Petruccione, "An Introduction to Quantum Machine Learning Algorithms", Contemporary Physics, vol. 56, no. 1366-5812, pp. 172-185, 2015.

[6] Christopher Havenstein, Damarcus Thomas and Swami Chandrasekaran, "Comparisons of performance between quantum and classical machine learning", SMU Data Science Review, vol. 1, no. 4, pp. 11, 2018.

[7] Carmen G Almudever, Lingling Lao, Robert Wille and Gian G Guerreschi, "Realizing quantum algorithms on real quantum computing devices", 2020 Design Automation & Test in Europe Conference & Exhibition (DATE), pp. 864-872, 2020.

[8] Sayantan Gupta, Subhrodip Mohanta, Mayukh Chakraborty and Souradeep Ghosh, "Quantum machine learning-using quantum computation in artificial intelligence and deep neural networks: Quantum computation and machine learning in artificial intelligence", 2017 8th Annual Industrial Automation and Electromechanical Engineering Conference (IEMECON), pp. 268-274, 2017.

[9] Yao Zhang and Qiang Ni, "Recent advances in quantum machine learning", Quantum Engineering, vol. 2, no. 1, 2020.

[10] Leonardo Alchieri, Davide Badalotti, Pietro Bonardi, Simone Bianco," An introduction to quantum machine learning: from quantum logic to quantum deep learning", Journal of Quantum Machine Intelligence, Vol 3 (2), 2021, 10.1007/s42484-021-00056-8.

[11] Vandana Rohokale, Milind Rohokale, "Quantum Computing: The Essence of Next Generation Smart Networks", Journal of the Maharaja

Sayajirao University of Baroda, Vol. 55, No.1 (IV), ISSN: 0025-0422, 2021.

[12] Antonio Manzalini, "Quantum Communications in Future Networks and Services", MDPI Quantum Report, March 2020.

[13] Dmitry Solenov, Jay Brieler, Jeffrey F. Scherrer, "The Potential of Quantum Computing and Machine Learning to Advance Clinical Research and Change the Practice of Medicine", Missouri Medicine, September/October 2018.

[14] Gulshan V, Peng L, Coram M, et al. "Development and Validation of a Deep Learning Algorithm for Detection of Diabetic Retinopathy in Retinal Fundus Photographs", 2016.

[15] Phillipson, Frank, "Quantum Machine Learning: Benefits and Practical Example", 2020.

[16] Stephen Gosste, "10 Quantum Computing Applications and Examples", Blog in Builtin, June 2022, https://builtin.com/hardware/quantumcomputing-applications

13

Unlocking Inner Potential: Tactile Internet's Impact on Healing and Self-actualization

Swati Sharma[1] and Ramjee Prasad[2]

[1]Vishwaniketan Institute of Management, Entrepreneurship & Technology, India
[2]CTIF Global Capsule (CGC) Herning, Denmark
E-mail: swati.mindscuba@gmail.com; prasad.ramjee@ctifglobalcapsule.org

Abstract

This chapter delves into the profound potential of inner work, self-awareness, and time-honored techniques in the pursuit of attaining C5 – an encompassing state comprising concentration, confidence, creativity, calmness, and contentment. The ultimate goal is to facilitate not only profound healing but also a substantial enhancement in day-to-day quality of life (QoL).

Furthermore, we shed light on the pivotal role that wireless tactile technology can play in conducting C5 practices, offering a personal touch without being confined by geographical or temporal constraints. By synergizing the potency of personal growth practices with the opportunities presented by advancements in tactile internet, individuals can embark on a transformative journey toward realizing the state of C5 and achieving a deeper sense of overall well-being.

It is crucial to underscore that the primary focus of this chapter lies in the conceptualization phase, specifically exploring how the tactile internet can serve as a platform for delivering C5 coaching.

5G and 6G deal with broad technology, particularly 6G, which covers the aspect of terrestrial communication, as well as space and satellite communication. Looking at the point of C5, it deals with the applications of human beings at the terrestrial level or space and satellite level. Therefore, this chapter is of great interest for 5G/6G applications.

DOI: 10.1201/9788743801108-15

Our aim in the current chapter is to lay the groundwork for future research endeavors, with an eye toward experimentation and in-depth investigation. Subsequent phases will see us delving into these critical aspects.

Keywords: Inner potential, tactile internet, five elements of nature, air, water, fire, earth, space, haptic devices, wearables

13.1 Introduction

In the relentless pace of our modern lives, it is commonplace to find ourselves caught in the whirlwind of routine, operating on autopilot as we navigate through the incessant demands of tasks, vacations, and hobbies. Yet, these pursuits often become mere checkboxes on our to-do lists, devoid of genuine moments of disconnection. In our pursuit of constant activity, we neglect the essential need for pauses, moments where our bodies can rest from the ceaseless chatter of our overactive minds.

Ancient Indian philosophies impart a profound perspective, suggesting that our mind constitutes our subtle body, while our body embodies our gross mind. Unfortunately, we have allowed our minds to overpower us, drowning out the voices of our bodies. The fear of inactivity compels us to incessantly stimulate our minds, perpetuating a cycle that leads to anxiety and a relentless craving for quick dopamine hits.

From the crack of dawn to the moment we retire to bed, we remain tethered to our smartphones, these digital companions evolving into extensions of ourselves. This constant connectivity leaves us with minds overstimulated and bodies utterly fatigued. Chronic exhaustion, lethargy, and a pervasive sense of emptiness become the unwelcome companions of our daily existence.

Yet, the challenges do not end there. We carry emotional baggage – past wounds, traumas, limiting beliefs, and victim mentalities – that can surface unexpectedly, triggering us to seek solace in temporary comforts like food, alcohol, or the relentless scroll of social media. As Dumbledore wisely noted, "Numbing the pain for a while will make it worse when you finally feel it" [1].

In the face of these trials, hope prevails. Timeless philosophies and healing techniques beckon us to break free from this damaging pattern of existence. While the familiarity with these methods may linger, the invitation now is to trust the process wholeheartedly. These ancient practices possess the transformative power to heal us at the deepest levels, guiding us toward a life of fulfillment – an existence in harmony with the secret of C5.

The question posed is clear: are you prepared to embark on this transformative journey and discover the illumination at the end of the tunnel? The invitation stands – a beckoning call to reclaim a life envisioned, where the wisdom of the ages converges with the essence of C5 to illuminate the path toward profound well-being and fulfillment.

13.2 C5 – A Way to Holistic Living

What is C5 after all, and how does it hold the key to a wholehearted existence?

At its core, C5 [2], as introduced by the author of "Unlock Your Personalization," stands as a transformative concept that encapsulates a holistic approach to life.

Crafted through years of dedicated work and personally internalized by the author, C5 has become a guiding principle that he has shared with individuals worldwide.

The essence of C5 is deeply rooted in the belief that possessing each of the Cs, unlocks the gateway to a life marked by happiness, dynamism, and whole-heartedness.

C5 stands for
Concentration
Confidence
Creativity
Calmness
Contentment

This concept transcends a mere collection of qualities; it represents a philosophy that, when embraced, has the potential to revolutionize how one engages with life. It signifies more than a checklist of attributes; it is a blueprint for a life lived with intention and fulfillment. The journey toward a whole-hearted existence, according to C5, involves cultivating these fundamental elements within oneself, paving the way for a dynamic and joyous life. Through the acquisition and nurturing of concentration, confidence, creativity, calmness, and contentment, individuals can unlock their true potential and experience the richness of a life well-lived.

As illustrated in Figure 13.1, all 5Cs are interrelated, and how each of their presence can make the others flow.

Let us delve into a detailed exploration of the interrelationship among the 5Cs – concentration, confidence, creativity, calmness, and contentment – and how they contribute to our overall well-being.

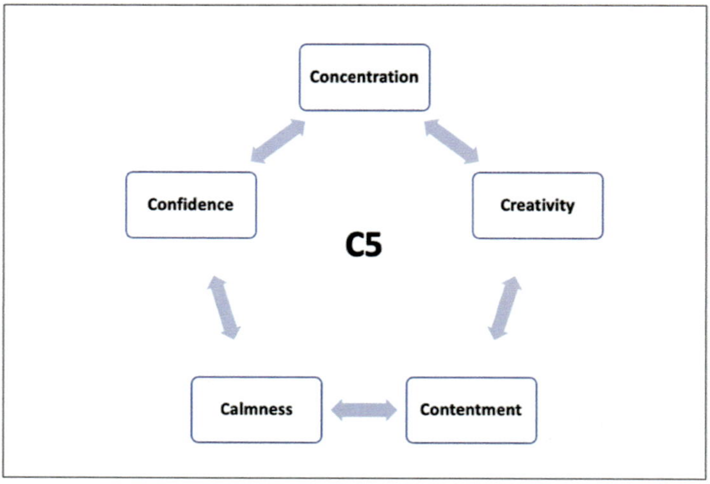

Figure 13.1 C5 concept.

Interconnectedness of the 5Cs:

Concentration, confidence, and creativity:

Concentration acts as a bridge between confidence and creativity. When concentration is high, it enhances both confidence and creativity. On the flip side, low confidence can impede concentration, hindering the expression of creativity.

Confidence and creativity are intertwined; a lack of confidence may stifle creative expression, while the exploration of one's creativity can boost confidence.

Confidence and creativity with calmness and contentment:

Confidence contributes to calmness, as a secure sense of self, fosters inner tranquility. Simultaneously, calmness supports the growth of confidence by providing a stable emotional foundation.

Creativity, when coupled with contentment, allows for a harmonious expression of one's ideas and innovations. Contentment, in turn, is influenced positively by the act of creating and expressing oneself.

Foundation of well-being:

The 5Cs collectively form the core of our well-being. They create a dynamic equilibrium, and any imbalance in one may affect the others, leading to a sense of incompleteness.

Let us understand this further by the following case studies.

Case study 1:

Joanna is a 32-year-old woman constantly bursting with ideas and a deep desire to share her creativity with the world. However, she lives in constant self-doubt and is unable to muster enough confidence to share her ideas with the world. What if they are not good enough, what if others will make fun of her! This constant doubting and questioning acts as a barrier, hindering her from fully expressing her creativity. This imbalance has a direct cascading effect, naturally diminishing the other Cs – concentration, calmness, and contentment in her life.

Addressing Joanna's lack of confidence becomes crucial to restoring the equilibrium of the 5Cs in her life. This would need inner healing and also inviting activities and elements that can directly improve her self-confidence.

Case study 2:

Now, let us turn our attention to *Ryan*, a 40-year-old man, who has a stable job and a general sense of contentment with his life situation. However, he feels a void because he is unable to find an outlet for his creative energies. He feels he is stuck in a rut with no end to come out of it. Without a means to channel his creative juices, there will come a point where Ryan will start to experience a decline in contentment, calmness, concentration, and finally, confidence.

Encouraging Ryan to find a creative outlet is pivotal for re-establishing the balance of the 5Cs in his life.

As can be observed, even the absence of just one of these Cs can create a sense of imbalance within us. We may feel that something is not quite right, as if an essential piece of the puzzle is missing. It is through nurturing and cultivating all of these Cs that we can unlock our true potential and experience a deep sense of harmony within ourselves and in our lives.

Embarking on the transformative journey of cultivating the 5Cs – concentration, confidence, creativity, calmness, and contentment – unveils the pathway to unlock our unique personalization. This profound expedition leads us toward the pinnacle of self-discovery and fulfillment, a state often referred to as "self-actualization."

Historically, self-actualization held the coveted position as the zenith of human needs in Maslow's well-known hierarchy [3]. However, as the esteemed American psychologist, Abraham Maslow, refined his model, a new summit emerged – self-transcendence. This evolution marked a paradigm shift in understanding our innate aspirations.

As we nurture the 5Cs within ourselves, we venture into a realm where personalization becomes more than a concept; it transforms into a lived experience. This journey is a dynamic process of self-realization, where each "C" contributes uniquely to our holistic well-being. The harmonious interplay of concentration, confidence, creativity, calmness, and contentment becomes the canvas upon which we paint our individual narratives.

This quest for personalization is not merely a pursuit of happiness but an expedition toward living our best life, embracing our full potential, and ultimately reaching the state of self-actualization. It signifies a profound connection with one's authentic self and the realization of intrinsic capabilities.

In acknowledging Maslow's evolution from self-actualization to self-transcendence, we recognize that our journey is not solitary; it extends beyond ourselves. The pursuit of self-transcendence involves reaching beyond personal interests and connecting with a larger purpose – a transcendence of the individual self for the greater good.

As we engage with the 5Cs and weave them into the fabric of our lives, we open the door to self-discovery, growth, and a profound sense of purpose. This transformative process not only aligns with Maslow's refined hierarchy but also reflects the ongoing evolution of our understanding of what it means to lead a truly fulfilling life.

Is there a way wherein these 5 Cs can intrinsically and intuitively become a part of or being?

This is where the methodology of the *five elements of nature* comes into play.

By aligning ourselves with the qualities represented by the elements – air, water, fire, earth, and space – we can cultivate and embody all the 5Cs within us.

13.3 Methodology

Let us explore how the five elements of nature can allow us to tap into the innate wisdom of nature and leverage its transformative power to develop concentration, confidence, creativity, calmness, and contentment (C5) as integral parts of our existence.

13.3.1 The five elements of nature philosophy

Our physical existence is intricately woven from the fabric of five fundamental elements: water, air, fire, earth, and space [4]. In the harmonious symphony of these elements, our bodies find equilibrium, resulting in a profound sense of vitality and overall well-being. It is the delicate balance of these elements that sustains our health.

However, the equilibrium is delicate, and any deviation, be it an imbalance or excess of a particular element, can manifest in various physical and mental disturbances. The synergy of these elements extends beyond the realms of physical health, exerting a direct influence on the nuanced aspects of our mental and emotional well-being.

The interplay between the five elements and the qualities of C5 – concentration, confidence, creativity, calmness, and contentment – is both fascinating and transformative. Recognizing this interconnectedness allows us to embark on a journey of self-discovery and intentional living.

To honor and nurture each element in our daily lives becomes a profound practice. It involves not only acknowledging the elemental composition of our being but also understanding how these elements shape and influence the subtle aspects of our inner selves. By doing so, we unlock the potential to cultivate and elevate the 5Cs within us.

In essence, the recognition of the interdependent relationship between the five elements and C5 empowers us to become custodians of our well-being. Through conscious choices, practices, and a deepened connection with the elements, we embark on a journey of holistic growth – nurturing both our physical health and the nuanced qualities that contribute to a fulfilling and harmonious life. Let us take a look at Figure 13.2 and explore the connection between each element and its associated Cs.

1. *Air* is linked to *concentration*
   Air symbolizes mental focus and clarity. Fresh air is often recommended to revitalize concentration and clarity. Delving deeper, the incorporation of practices involving deep, mindful breathing becomes a homage to the air element. Through intentional breathing exercises, we not only honor this elemental association but also enhance our cognitive abilities, fostering a heightened capacity to concentrate.
2. *Water* is associated with *confidence*
   Water, with its fluidity, adaptability, and inherent strength, serves as a metaphor for building confidence. Mimicking water's ability to flow and adapt to its surroundings, we can internalize these qualities to bolster our self-assurance and belief in ourselves. Drawing inspiration from the resilience of water, we learn to navigate life's challenges with grace and confidence.

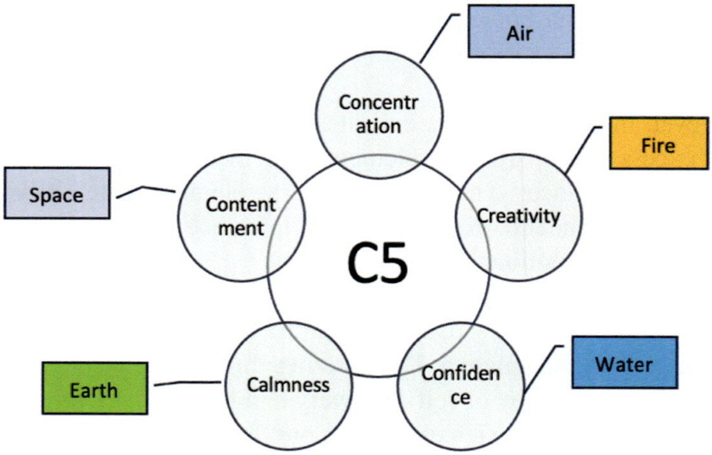

Figure 13.2 C5 and its association with five elements of nature.

3. *Fire* is connected to *creativity*:
   The transformative and dynamic nature of fire serves as a catalyst for creative thinking and passion. The phrase "fired up" directly acknowledges the connection between fire and creativity. By igniting our internal fire element, we tap into a wellspring of creative energy and potential. Embracing the spark of fire within us becomes a means to foster innovation and artistic expression.
4. *Earth* is associated with *calmness*:
   The earth element embodies stability, grounding, and nurturing characteristics that contribute to a profound sense of calmness and stability. Grounding ourselves in the present moment and connecting with nature allows us to cultivate this state of calmness and inner peace. By consciously engaging with the earth element, we anchor ourselves in a serene foundation, fostering emotional equilibrium.
5. *Space* is linked to *contentment*:
   Space, representing openness, expansiveness, and freedom, holds the key to contentment. Creating physical and mental space in our lives becomes a practice of allowing room for contentment and peace to flourish. Meditation, as a transformative practice, aligns with the expansive nature of space, offering a gateway to infinite possibilities and a profound sense of contentment. Through intentional space creation, we invite tranquility into our lives, embracing the boundless potential for inner fulfillment.

13.3.2 Honoring five elements of nature in our daily life

While the demands of our hectic schedules may make it seem challenging to honor each of the five elements in our daily lives, incorporating these practices does not have to be an arduous task. What if, like the daily routines of brushing and bathing, these elemental acknowledgments could seamlessly become a natural part of our day? Let us explore below how this integration can be achieved through simple and basic activities

1. *Water* element: starting your day with a warm glass of water. Begin your day by hydrating your body. For enhanced benefits, you can make it alkaline by adding a few drops of lemon juice or a spoonful of apple cider vinegar. Although lemon and apple cider vinegar are acidic in nature, they have an alkalizing effect when metabolized, thereby raising the pH levels in our bodies. This practice honors the *water* element and replenishes our bodies (which is approx. 60% water) right at the onset of the day.
2. *Fire* element:
    - Invoking the higher power and practicing Sungazing [5]. Take a moment to connect with a higher power of the divine. We can honor the *fire* element by engaging in Sungazing, which involves gazing at the sun for 5 minutes *only* during safe hours (dawn or dusk). The sun serves as our life-giving energy source, and this practice acknowledges and appreciates its vital role.
    - *Fire* also represents energy, strength, and transformation. Honor this element by incorporating physical exercise into your routine. Whether it is cardio, strength training, or any other form of physical activity, it helps invigorate the *fire* within you, promoting vitality and overall well-being.
3. *Air* element: practicing pranayama, breathing exercises. *Air* symbolizes breath and life force energy. By incorporating pranayama or breathing exercises into your daily practice, you can regulate and deepen your breath, promoting a sense of calm, clarity, and improved respiratory function. By focusing on the air element, you can enhance your connection with your breath and cultivate concentration.
4. *Earth* element: incorporating relaxation techniques, such as shavasana. *Earth* represents stability, grounding, and relaxation. After physical exertion or at the end of your morning practice, embrace relaxation techniques like shavasana (corpse pose) in yoga. This posture allows

you to rest and surrender, connecting with the earth's nurturing energy and promoting a sense of tranquility.

5. *Space* element: engaging in meditation and centering exercises. *Space* represents vastness, clarity, and expansiveness. Meditation and centering practices cultivate mindfulness, focus, and inner stillness, allowing you to expand your awareness and tap into a state of profound presence leading to deeper contentment.

Each of these activities has deeper layers and the correct ways of doing them. To initiate or start these activities, one would need the help of a practiced C5 coach who can guide you through each step properly and also explain the importance of each of these steps in a detailed and scientific manner.

Please note: barring points 1 and 2 above, others can be done at any time of the day provided the preconditions for them are met.

The first key step to align the elements and C5 is

- *Mindfulness*: cultivate awareness of each element's qualities within yourself.
- *Practices*: engage in activities that connect you with nature, such as meditation, spending time outdoors, or incorporating elemental symbolism into your surroundings.

And with the right intention and practices in place, we will be on the path of Integrating C5 into our being:

- Airy mindset: embrace clarity and focus in your thoughts and actions.
- Fluid confidence: adapt to challenges with resilience and fluidity.
- Passionate creativity: channel inner fire into creative pursuits.
- Grounded calmness: establish stability amidst life's fluctuations.
- Spacious contentment: embrace openness and find contentment in the present moment.

Upon delving into the practices and principles outlined, if you find yourself inspired to embark on this transformative journey independently, feel free to do so with full enthusiasm. However, if you sense that the guidance of a seasoned teacher or coach could enhance your experience, then the C5 coach is the ideal solution. This specialized coach is not only well-versed in the intricacies of the 5Cs and the five elements but is also equipped to offer guidance on preparing both your body and mind for the activities [6].

The role of the C5 coach extends beyond conventional coaching. This professional is not only certified in modern technologies and philosophies but is deeply rooted in ancient Indian wisdom. Such a coach is uniquely positioned to impart training on incorporating the 5Cs and the five elements into your daily life. Furthermore, the coach will be adept at providing support for emotional issues and personal challenges, drawing from a holistic perspective that integrates both ancient wisdom and contemporary methodologies.

Locating individuals with this combination of modern insight and deep traditional knowledge may seem challenging. The envisioned C5 coach embodies a harmonious blend of modernity and tradition, making them a rare find in the coaching landscape. The objective is to extend the impact of these coaches to a broader audience, transcending the constraints of time and location.

To overcome the challenges posed by geographical and temporal limitations, our strategy involves integrating tactile internet and haptic devices into our practice. This forward-thinking approach aims to make the expertise of C5 coaches more accessible to individuals regardless of their location or time zone. By leveraging these technologies, we seek to maximize the reach of these coaches, ensuring that transformative guidance is readily available to anyone seeking to integrate the principles of C5 and the five elements into their lives.

13.4 Tactile Internet and C5

The continuous evolution of wireless communications technology, coupled with the emergence of the tactile internet [7], fuels our imagination for a future unbound by geographical constraints and time limitations. Envision a world where the barriers of distance and time no longer impede our ability to connect and communicate.

In the domain of healing and personal growth, these technological strides hold the key to profound transformations. They open doors for C5 coaches to extend their guidance, support, and that invaluable sense of connection and touch, even within the virtual realm. This unmatched potential of the tactile internet in facilitating healing and nurturing through the 5Cs through online coaching is undeniably awe-inspiring.

The tactile internet represents a paradigm shift, providing an advanced level of connectivity that transcends the limitations of traditional online interactions. Imagine a scenario where a C5 coach, equipped with the tools enabled by the tactile internet, can virtually provide the kind of support that was once only possible through in-person sessions. The sense of touch, often considered irreplaceable, becomes tangible even in the virtual landscape.

This technological leap opens new horizons for healing practices and personal growth, making them accessible to individuals irrespective of their physical location. It introduces a dynamic where the virtual realm becomes a powerful space for profound connections, emotional support, and transformative coaching experiences. The potential impact of the tactile internet on the field of healing and the cultivation of the 5Cs is not just revolutionary; it is a testament to the boundless possibilities that technology can offer in enhancing our well-being and personal development.

As described in Figure 13.3, the integration of tactile technology and haptic devices is envisioned at both the client and coach ends. This strategic placement ensures a dynamic exchange of sensory data, crafting a personalized and immersive experience tailored to each client. The bidirectional flow of tactile information contributes to an environment where interactions feel not only personal but also finely customized to meet the unique needs and preferences of every individual.

Let us explore how this ground-breaking technology can bring a personal touch and enhance the field of C5 coaching conducted online.

- *Nonverbal communication*
  We know that words do not always capture the full essence of our thoughts and emotions. The tactile internet introduces the potential for C5 coaches to utilize haptic feedback devices or wearables. By incorporating touch, gestures, and body language into online sessions, coaches can establish a richer connection with clients. This integration enhances understanding and trust, leveraging nonverbal cues for a more profound coaching experience.
- *Emotional support*
  Emotional well-being is central to C5 coaching. With the tactile internet, coaches can use haptic devices to provide comforting touch sensations from a distance. Imagine feeling a reassuring touch during vulnerable

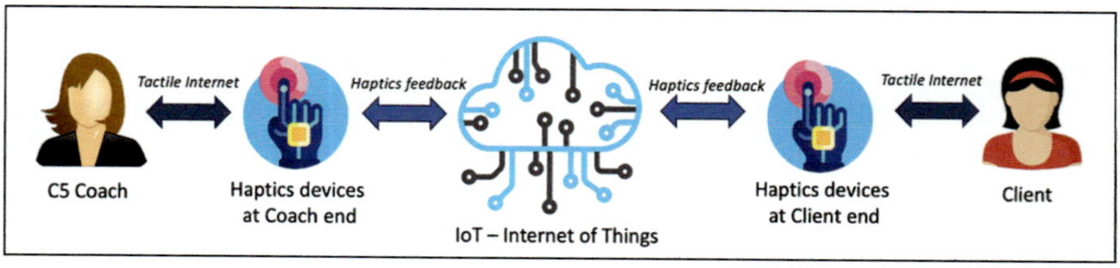

Figure 13.3 C5 coaching using tactile internet.

moments or receiving a virtual hug when needed. These technological innovations foster trust, empathy, and a genuine emotional bond between coach and client, transcending physical distance.

- *Experiential learning*
  We learn best through experience [8]. The tactile internet opens up a world of possibilities for experiential learning in online coaching sessions. Clients can engage in interactive simulations or virtual reality experiences that incorporate tactile feedback. Through these immersive activities, guided by their C5 coach, clients can practice and develop skills in a tangible and deeply engaging way.
- *Boundary setting*
  Establishing and respecting personal boundaries is crucial in any coaching journey. Respecting one's own and other's boundaries is one of the most important aspects of life coaching. The tactile internet offers a unique opportunity for coaches and clients to explore and practice boundary-setting exercises. By using haptic feedback devices, clients can experience the sensation of physical boundaries, helping them understand and assert their personal space and limits with confidence.
- *Kinesthetic learning*
  Each person has a unique learning style, and for some, physical movement and kinesthetic experiences are vital for growth [9]. With the tactile internet, C5 coaches can integrate haptic feedback devices or wearable technologies that provide tactile cues during kinesthetic activities or exercises. For instance, take the breathing exercise or pranayama, this can aid in helping the client understand the right way of doing it which sometimes cannot be learned just by watching a video.
- *Remote support*:
  The beauty of the tactile internet lies in its ability to extend support to individuals who may be physically distant or face mobility challenges. It eliminates the barriers of time and location, allowing C5 coaches to reach a wider audience. Now, more than ever, individuals from various corners of the world can access the guidance and support they deserve, enriching their lives with the transformative power of the C5 concept.

In essence, the tactile internet unfolds a realm of possibilities, elevating the personal connection and transformative potential of online life coaching. It erases physical barriers, delivering empathy, guidance, and the human touch crucial to personal growth and well-being. This exciting future, where technology and empathy converge, invites us on a journey of self-discovery and empowerment together. Let us embrace this convergence and embark on a path of holistic transformation.

13.5 Discussion and Future Scope

The future trajectory of haptic technology looks promising with several key points to consider.

- *Availability and interfacing*: accessibility and interface ease will become the defining features of haptic technology. This trend suggests a growing adoption across various domains, ranging from gaming to health care and education. The increasing availability of haptic devices makes it more feasible for users to integrate touch-based interactions into their daily lives.
- *Reliability of sensory data*: the reliability of sensory data gathered by haptic devices is of paramount importance. Ensuring accuracy and consistency in the feedback provided by these devices is a critical focus of ongoing research and development efforts. This emphasis on data reliability is crucial for the effectiveness of haptic technology in delivering immersive and responsive user experiences.
- *Advancement in haptic technology*: anticipate a continuous wave of advancements in haptic technology. These developments are expected to yield more sophisticated haptic interfaces, enhancing feedback precision and expanding the capacity to simulate a broader range of tactile sensations. This evolution holds significant implications for applications in virtual reality, gaming, and beyond, where nuanced touch experiences are increasingly valued.
- *Affordability and accessibility*: as haptic technology matures and economies of scale come into play, a notable trend is the increased affordability and accessibility of haptic devices. This shift makes these devices more attainable for a broader user base, opening up new possibilities for applications in diverse fields such as gaming, virtual reality, health care, education, and even, innovative coaching approaches like C5 coaching.

In summary, the future of haptic technology appears poised for substantial growth and development. Its increasing availability, enhanced reliability, continuous technological advancements, and improved affordability collectively contribute to the unfolding landscape of innovative applications. From enriching gaming experiences to revolutionizing health care and educational interactions, haptic technology is set to redefine how we engage with the digital world, emphasizing the power of touch and physical feedback.

13.6 Conclusions

To sum it up, the 5Cs – concentration, confidence, creativity, calmness, and contentment – are essential for a fulfilling life. Integrating these principles into our daily lives leads us toward a meaningful and fulfilling existence.

The elements of earth, air, water, fire, and space symbolize the foundation of the C5 approach. Harmonizing with these elements unlocks C5 and this in turn unlocks our potential, our personalization.

In this day and age of technological advancement, let us take it into our stride and use it to elevate our entire experience. Tactile internet and haptic feedback can play a transformative role in embracing the C5 way of living. Haptic technology can become a catalyst for fostering human connection and allowing us to connect deeply with our surroundings.

Haptic technology can make the C5 principles accessible globally, enhancing human–computer interaction and bridging the virtual and physical worlds.

Imagine a world where technology feels warm and connected, making whole-hearted living natural for everyone. That is our goal – bringing back the human touch to our tech-filled lives.

Looking ahead, we are diving into some serious research. This chapter will be further expanded for future research in the area of application and experiments.

We will figure out how all systems will work together and draw detailed block diagrams along with system specifications for ease of understanding and practical applications of our concept. This will help us take our C5 ideas and technology to the next level, making it simple for everyone to embrace a fulfilling and connected way of living. It is like setting the stage for a technology-powered and yet warm-hearted world.

References

[1] Harry Potter and the Goblet of Fire, J.K. Rowling, *2000.*
[2] Ramjee Prasad "Unlock your personalization", Aalborg Universitetsforlag, *2012.*
[3] Maslow's Hierarchy of Needs, Saul McLeod, *May 2018.*

[4] Nature's five elements: Connection with the human body, their role in human life and wellness by C D Marathe & Jagdish Acharya (*June 2020*).
[5] Active Transformation of mind and body through abundant natural methods (ATMAN), Shobha Lal, Ambuj Kumar, Ramjee Prasad, *2019.*
[6] A Pragmatic Business Approach to a Novel C5 Concept ATMAN by Shobha Lala, Ramjee Prasad, *2021.*
[7] C5: A Step towards Smart World with Enhanced Holistic Wellbeing by Smita Chandra, Neeli Rashmi Prasad, Peter Lindgren, and Ramjee Prasad, *2022.*
[8] Haptic Technology by Mohan N Pujari, Dr Vishal C (*July 2022*).
[9] Concept and Application of Virtual Reality Haptic Technology: a review by Nuratiqa Natrah Mansor, Muhammad Herman Jamaluddin, Ahmad Zaki Shukor (*July 2017*).

14

Efficient Communication for Real-time Systems

Ramesh Yadava[1] and Ramjee Prasad[2]

[1]Department of Production and Industrial Engineering, BIT Sindri, India
[2]CTIF Global Capsule (CGC), Herning, Denmark
E-mail: rameshyadava@bitsindri.ac.in; prasad.ramjee@ctifglobalcapsule.org

Abstract

This chapter addresses the limited bandwidth challenge that is a bottleneck for real-time data transfer. The COVID-19 pandemic has dramatically changed how we work in businesses and at home. Remote work has become increasingly common, leading to a global surge in internet and bandwidth usage. This growing demand extends beyond just video conferencing. The future of zero-emission transportation (electric vehicles), artificial intelligence (AI), edge computing (EC), software-defined networking (SDN), network functional visualization (NFV), and ubiquitous data collection across sectors like education (remote learning), health care (telemedicine), and even food delivery – all rely on a robust and fast internet connection. To meet this need, wireless communication is evolving toward 6G and beyond. However, the pace of adoption varies greatly between countries due to economic and other factors. This creates a significant "bandwidth gap" – a stark difference in internet speeds across the globe. Here is the key challenge: as technology advances toward 6G, this gap will likely persist. In any real-time data transfer scenario, the client with the lowest bandwidth acts as a bottleneck, hindering the overall efficiency of everyone involved. This is particularly evident in video conferencing, where participants join from various locations with potentially vastly different internet speeds. The result is choppy video, frozen presentations, lost packets, increased latency (delay), and even dropped connections. All of these translate to lost productivity

DOI: 10.1201/9788743801108-16

and frustration. This research aims to develop a solution that ensures clients with limited bandwidth do not drag down the efficiency of the system and negatively impact the experience for participants with higher bandwidth capabilities.

Keywords: Low bandwidth, video conference system, video conference call, QoS, QoE, edge computing, software-defined network, network functional visualization

14.1 Introduction

In the video conference meeting, the user with the lowest bandwidth brings down the efficiency of the entire system, holding everyone waiting for video and audio to synchronize and stabilize. The exploding requirements (Figure 14.1) of sensors and other applications in various daily-use equipment have increased the bandwidth requirement. To address this need, the industry cellular network has migrated from 1G → 2G → 3G → 4G → 5G [1], and soon it will be 6G and beyond. The use of the internet and the bandwidth have grown exponentially [2] all across the globe. Zero-emission transport, that is, future electronic vehicle (EV) solutions, is starting to have a place in the industry. Additionally, it is important to note the growth in artificial intelligence (AI) for applications such as edge computing (EC) and software-defined networking (SDN) in every aspect of our lives, from health and food to lifestyle requiring connectivity. This has demanded a drastic increase in Internet speed. To address this need, wireless communication is entering into the next generations of 6G and beyond, as listed in Table 14.1, and the forecast and global trends by year are listed in Table 14.2.

The Internet of Things (IoT) is a network of physical devices that uses sensors or software to collect data and pass it for data computation. EC technology moves data computation and storage closer to data sources such as IoT. This technology not only improves response time but also reduces latency and bandwidth requirements. SDN uses software-based controllers or application programming interfaces (API). SDN is promoted by the Open Network Foundation (ONF). This technology makes it easier to build and manage large complex networks. SDN traffic is managed from a centralized control point by changing the rules for any switch in the system as required. Network functional visualization (NFV) uses virtualization technologies to separate the network functions from hardware into software, enabling the

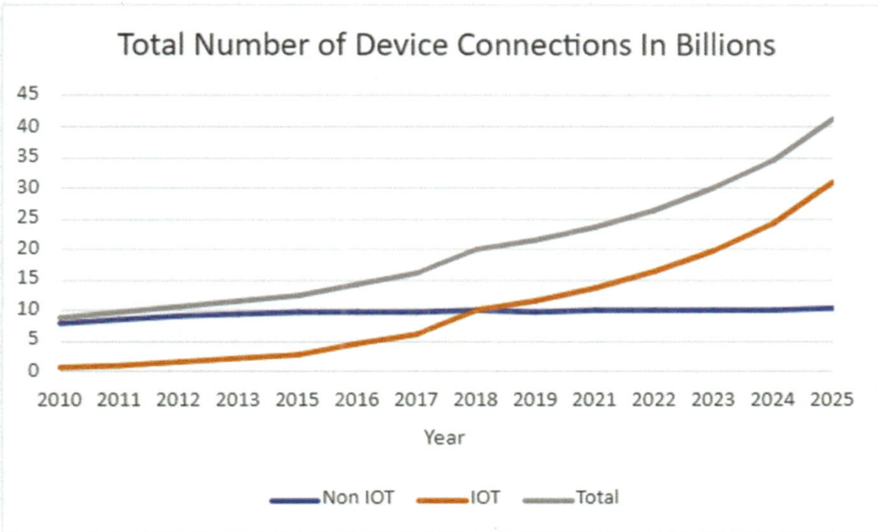

Figure 14.1 Estimated connected devices [5].

Table 14.1 History of wireless communication technology growth 1G-5G [3], 6G projected implementation [4].

Generation	Year	Description
1G (First generation)	1980's	First mobile phone application (Motorola)
2G	1990's	Short Message Service, (SMS) and roaming
3G	2000's	Smart Phone (Apple I-Phone) enabling Internet
4G	2010's	Increased Bandwidth enables real time applications, Higher accuracy in GPS, Vehicle auto control
5G	2020's	Increased Bandwidth speed opened opportunity to make interconnection between various systems and sensors changing the lifestyle for better
6G	2030's	With superfast Bandwidth, the connectivity and control among various systems (interdisciplinary) in space, ground and underwater is a reality

Table 14.2 Forecast and trends global by year [6].

	2017	**2022**
Internet Users % of Population	45.30%	59.70%
Devices and Connection per Capita	2.4	3.6
Average Speeds	39 Mbps	75.4 Mbps
Average Speed per Capita/month	16.2 GB	49.8 GB

service provider to run their network on standard servers instead of custom ones. NFV is promoted by the European Telecommunications Standards Institute (ETSI). NFV increases telecom network resource utilization with the added advantage of reducing the cost of developing and implementing the solution.

There are 195 countries, and as of December 2022, 72 countries [10] have implemented or are in the process of implementing 5G. Since 5G is up to 100 times faster than 4G, there is a quantum leap in bandwidth availability when migrating to 5G and above. There is a cost to migrate to the next level of cellular networks, and some of the countries are still on 3G cellular networks (Figure 14.2). This difference in the implementation of cellular network bandwidth will always exist, as evolving technologies keep increasing the efficiency and performance of cellular network bandwidth while older technology is still in use (Tables 14.3 and 14.4). 5G provides ultralow latency and greater bandwidth for the use of new services such as AI for applications (EC and SDN) in e-health, connected vehicles, traffic systems, IoT, and advanced mobile cloud gaming.

Users using an application with lower bandwidth are the weakest link in the bandwidth chain of all online clients. This will negatively affect the

- The pace of adopting new technology varies between the countries
- Cost of implementing new technology, Obsoleting older equipment is a challenge
- Some countries still have 5 Mbps where as others are in 5 G implementation and will be racing fast to 6G

Advanced nations will always be way ahead of nations with less developed infrastructure

Figure 14.2 Why countries lag in implementation [7].

Table 14.3 1G to 6G network speed [8].

Network	Bandwidth	Avg. Speed	Unit	Year
1G	2	2	Kbps	1979
2G	364	40	Kbps	1991
3G	3	0.3	Mbps	2001
4G	100	25	Mbps	2009
5G	10	0.15	Gbps	2019
6G	1	TBD	Tbps	2030

Table 14.4 Global average internet speed [9].

	Internet Speed		
Year	**Kbits/sec**	**Mbits/sec**	**Gbits/sec**
1990	12		
2000	100		
2010		20	
2020		70	
2030		600	
2040		650	
2050			50

efficiency of that application system. This chapter highlights this cellular network bandwidth disparity and finds a solution to this challenge. The bandwidth limitation affects the performance of EC. Since EC data is processed locally, to take advantage of this, there must be a low latency time for the data to be transferred to the requester. Lower bandwidth will cause higher latency, limited data offloading, and slower response time. The proposed solution will help run the EC data transfer at its most optimized speed. The lower bandwidth will not be a barrier, and the data will be transferred from IoT at the system-optimized speed, helping reduce the latency time, response time, and data offload for applications requiring real-time data analysis, such as EV, medical field IoT applications, and other similar applications.

In SDN, the lower bandwidth will make the communication between the controller and switches slower. Bandwidth constraints will cause congestion because most open flow switches are equipped with limited flow-table space, which may become insufficient when there are many concurrent flows in the network, thus straining the controller resources. Big data can be translated into more bandwidth requirements, and for remote places, that will be a challenge. In NFV, the lower bandwidth will cause limited throughput for virtual network function (VNF) and increased latency, which will have challenges in VNF chaining. It will cause a delay in data transfer between chained VNFs. This chapter is the first of the series and focuses on the concept and definition of the challenge. The future book research will include a detailed technical discussion, architectural diagrams, and flow charts describing the approach, solution, and challenges.

This chapter will address the above challenge by developing an APP for

a. Creating custom communication pipes using screen mirroring in the cloud.
b. Creating a lower bandwidth client's screen in the cloud as a high-resolution screen.

c. The presentation slides and prerecorded video are loaded into the cloud for screen share.
d. Using API specific to the video conference system (VCS), for example, Zoom, MS Teams, Google Meet, Cisco Webex, Bluejeans, Neomeet, etc., EC, and SDN. The VCS, EC, and SDN are also regarded as hosts in this chapter.
e. Enabling "cloud-screen sharing" control using the client I/O devices (mouse/keyboard).

Since the client screen is in the cloud, the bandwidth limitation is not there anymore, allowing a client to present and display at higher resolution with other clients in the call.

The measurement of jitter, latency, and packet loss is in the future scope of this chapter.

14.2 Related Research: Issues and Challenges

This section highlights various industry solutions to provide high bandwidth across the globe for various resolution types and the solutions implanted to reduce latency.

In the video conference call, the user with the lowest bandwidth brings down the efficiency, causing everyone waiting for the video to stabilize and the audio to become clearer. This happens especially when a lower bandwidth client shares the screen and makes a slide deck presentation. In video conference applications (VCAs), the real time dynamic efficient spectrum analysis (RDEESA) technique is used for monitoring and analyzing the radio spectrum in real time. RDEESA increases audio quality by reducing noise, improving speech intelligibility, increasing video quality, and reducing the bandwidth requirement to transmit audio and video data. The VCAs measure the bandwidth of each system and accordingly adjust the video streaming rate (1080p, 720p, 480p, and 360p) as indicated in Table 14.5.

As of May 2023, over 190 countries are racing to install 5G cellular networks [12, 13]. Elon Musk has Starlink satellites [14] with an internet speed of 50 to 200 Mbps and covers 32 countries in North America, South America, Europe, Asia, Africa, and Australia. He plans to have satellites to cover the entire globe.

The Neomeet blog [15] describes how low bandwidth affects video conferencing and lowers quality of experience (QoE) and quality of service (QoS). Currently, there remains a significant difference in available

Table 14.5 List of video resolutions [11].

Resolution Type	Common Name p-Progressive video	Aspect Ratio	Pixel Size
SD (Standard Definition)	480 p	4 : 3	640 × 480
HD (High Definition)	720 p	16 : 9	1280 × 720
Full HD (FHD)	1080 p	16 : 9	1920 × 1080
QHD (Quad HD)	1440 p	16 : 9	2560 × 1440
2K video	1080 p	1 : 1.77	2048 × 1080
4 K video or Ultra HD (UHD)	4 K or 2160 p	1 : 1.9	3840 × 2160
8K video or Full Ultra HD	8 K or 4320 p	16 : 9	7680 × 4320

bandwidth (in various countries and clients) between the host and clients. The client with the lowest bandwidth cannot share the screen at the quality of HD 1080.

Networking companies, video conferencing companies, and business organizations have data centers with dedicated servers (in the cloud) to reduce the latency time, jitter, and packet loss and process video and data at the physically closest data center to the client. The difference in available bandwidth among the clients in video conferences hinders having an efficient video conference meeting.

As soon as the application "APP" determines the client's bandwidth is lower for video conference system use, it will create a virtual screen in the cloud and upload the "presentations data files" and "prerecorded video" to the cloud. There will be minimal change in the latency, delay, jitter, packet loss, or overhead, which will not affect performance.

14.3 Open Issues and Challenges

This section describes the solution to the lower bandwidth challenges.

- Develop a solution to create a custom communication pipe for each client based on (client's) bandwidth for a client to the host: the APP will determine the bandwidth of the client and set the policy for communication speed between the host and client, and if the available bandwidth is lower for video, the APP will create a virtual screen in the cloud and upload any presentation files as defined by the client.
- Develop a solution to create a virtual screen in the cloud for sharing the screen: the virtual screen in the cloud will enable a communication interface with the host at the highest bandwidth and video resolution.

- Develop a solution for virtual screen control by the client's keyboard and mouse: develop API for the client's keyboard and mouse to control the virtual screen.
- Video and audio synchronized: for QoE/QoS, the audio must be synchronized with the video.
- Develop a common API for the virtual screen in the cloud to communicate with the host.

Partner with VCS and businesses providing EC and SDN services for the development of API.

Cisco's Meraki white paper [6] identifies that the average monthly traffic per capita will increase from 16.2 GB per month in 2017 to 49.8 GB per month in 2022. The average projected global Wi-Fi network connection speed increased by 2.2× from 2017 to 2022.

The paper in [16] compares the performance of three VCA and determines that a temporary reduction in bandwidth may take as long as 50 seconds to recover to the steady state. The client utilization decreases as the number of participants increases due to reduced resolution depending on each participant's video stream. The underserved region experiences frequent connectivity disruption under dynamic network conditions. It talks about the negative effect of lower bandwidth but does not provide a solution.

The paper in [17] discusses a holistic multipath wireless video streaming survey. It describes the advantage of multipath communication in video processing to gain higher aggregate bandwidth. Besides, distributing video traffic over multiple network paths improves the client's stability, seamless connectivity, and QoE. This paper discusses various transmission protocols to improve the QoE. QoE depends on the available bandwidth for an application. This is addressed by adding multiple transmissions that deliver video streaming over IP broadcasts with bidirectional connectivity between the source and users. The challenges of multipath video transmission are being addressed aggressively in video streaming applications (protocol layer perspective: application layer, transport layer, network layer, and scheduling function to transmit video data over various wireless links, packet loss differentiation, fairness consideration, video CODEC, environment, performance metrics, video services, network heterogeneity, out-of-order packets, head-of-line [HOL] blocking, end-to-end delay, overdue packets, implementation aspects). Multipath transmission increases throughput, reliability, seamless connectivity, reduction of burst loss length, and delay improvement. The paper addresses the solution to video streaming. It does not address the VCA challenges of QoE and QoS for clients with lower bandwidth.

The paper in [18] reflects that cloud-based computing technology significantly improves video communication and networking, resulting in better QoE and QoS for video consumers. It examines the end-to-end partition of a cloud-based video system and evaluates its advantage in all major blocks, such as video server, content delivery network, and edge network with the cloud system, demonstrating the effectiveness of cloud computing.

For networked video services, currently, there are three critical issues: (1) limited computing and storage resources of thin client devices, (2) dynamic and heterogeneous network environment, and (3) mismatch between traditional QoS-based applications and QoE-driven video services.

Cloud computing technologies will dynamically allocate resources to satisfy the QoE and QoS expectations of the consumer. This will address the problem of thin-client challenges in video processing, addressing the issues listed in the above paragraph.

The paper emphasizes a balance between video QoE/QoS and the cost of computing and storage. The key challenge is to manage latency time, which affects the QoE and QoS. The paper identifies cloud-based video communication and networking challenges as service heterogeneity, terminal diversity, resource management, network dynamics, user experience, security, and privacy. The QoE and QoS will become even more challenging due to heterogeneous terminals and network modes. Since the video service is computation intensive and the thin clients have limited resources, the cloud service will be able to address both. The video is split into smaller packets and processed in parallel encoding, optimizing the encoding speed without compromising the video quality. In interactive applications, the total delay is the accumulation of the network, processing, and playout delays. For the video display, transcoding action is performed to match the bandwidth availability, increasing the QoE and QoS. Cloud computing technology provides the resources for thin clients to optimize the QoE/QoS, significantly increasing the video experience. This paper addresses mirroring the screen in the cloud. The cloud-based video communication and network distributes the load in the video server, content distribution network, and edge access network. Mirroring the screen of the client with lower bandwidth is still a lower resolution in mirrored screen. The paper addresses video streaming but does not give solutions for clients with lower bandwidth.

The article by Utility One expresses the role of QoE in video conferencing traffic management [19] and highlights the importance of QoE. Video conferencing is becoming an essential and required tool for businesses and teams to collaborate and communicate effectively across geographies to create the most efficient work environment to achieve the financial goal of profitability. The QoE and service play a vital role in video conferencing traffic management, impacting the effectiveness and productivity of these virtual meetings. This article identifies how businesses can optimize video conferencing systems for an enhanced user experience. Poor QoE during a video conference leads to frustration, miscommunications, and reduced productivity.

Report [20] has shown that video conferencing improves the efficiency and productivity of teams. On the other hand, the lower QoE will result in productivity loss, negatively affecting the bottom line. Lower bandwidth plays a major role in lower QoE.

The purpose of this chapter is to improve the QoE and QoS of clients with lower bandwidth by creating a screen in the cloud, breaking the shackles of bandwidth, and increasing the QoE and QoS. The author will expand the solution for EC and SDN in future publications.

14.4 Solution

Figure 14.3 demonstrates the suggested solution. The detailed description is provided below.

The host determines the bandwidth capability of each client and sets up the video streaming speed to communicate with this client without affecting the video streaming speed of all the other clients in the video call as shown in Figure 14.3.

1. A client communicates with the host at the client's bandwidth.
2. The final solution will determine the bandwidth of each "online client," and the host communicates with each client at the client's bandwidth.
3. The lower bandwidth client has all the data/video content uploaded to the cloud.
4. A virtual HD video screen is generated for clients (with lower bandwidth) in the cloud.
5. The audio signal is streamed to the virtual screen to synchronize with video.
6. Virtual screen control is from the client's I/O hardware.
7. Command to the VCS is controlled by client I/O hardware through a virtual screen.
8. The API will enable video feed from the virtual screen in the cloud to a VCS.

9. VCS will stream the video at the determined bandwidth as defined in item 1.

App solution: the video conference system (VCS) is the host.

1. The real-time applications APP being launched (Zoom, Skype, Webex, Teams, etc.)
2. The bandwidth of each client C_1 to C_N is determined by the host as $f(C_1)$ to $f(C_n)$ in Figure 14.4.
3. The host keeps tracking/recording the bandwidth of each client.
4. The host creates a custom communication bandwidth pipe for each client.
5. The host adjusts the resolution of the display to match the client's bandwidth and sets the rules/policies for communication between clients and the host.
6. The solution will be adopted for any application.
7. Multiple (low bandwidth) clients' virtual screens are in the cloud.

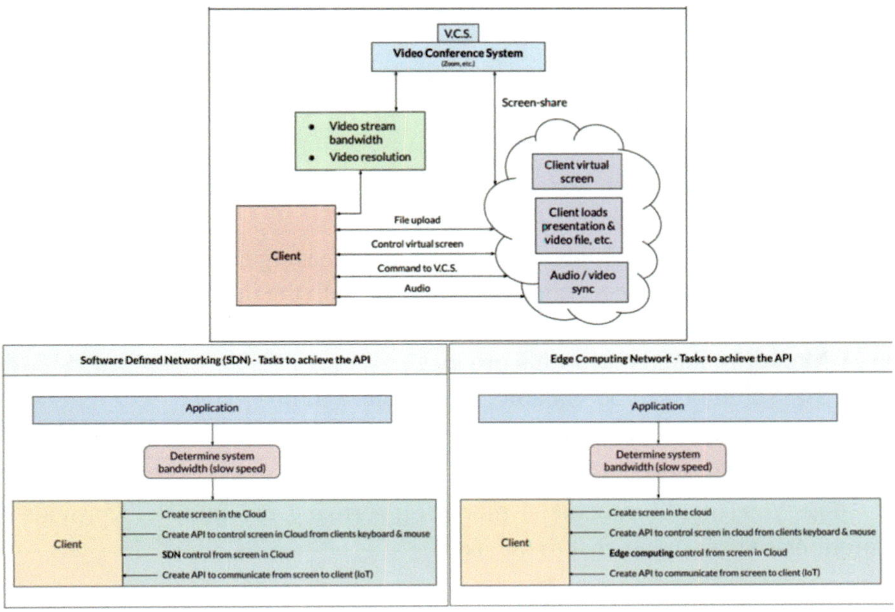

Figure 14.3 Efficient communication and screen sharing independent of the bandwidth of systems.

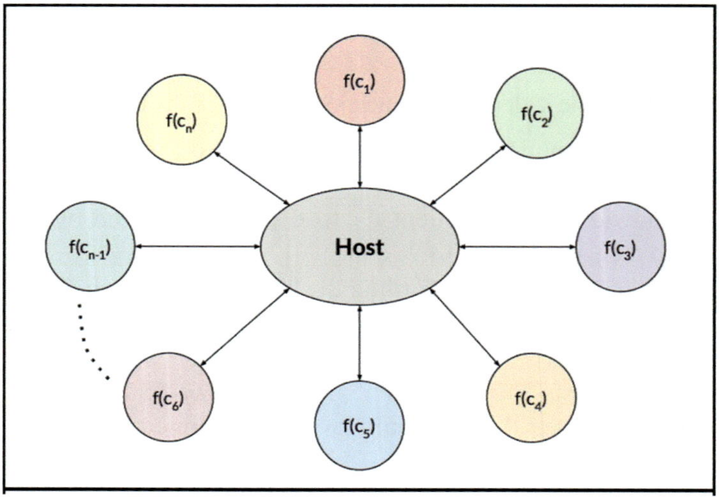

Figure 14.4 Communication pipeline at each individual client's bandwidth.

14.5 Discussion: Why This Solution? Potential Significance and Applications

The proposed solution will allow the system to operate efficiently and will not be affected by lower bandwidth clients. All clients online will communicate with the host at their bandwidth speed.

- The hiccups in the video conference calls rank amongst the highest complaints from users.
- The demand for this product will keep increasing as the communication bandwidth speed gets higher and higher, with the introduction of 6G and beyond.
- Market is abundant for this product.
- The solution will be scalable, repeatable, and profitable.
- The increase in efficiency and productivity of video conference meetings will offset the cost of the solution.

The YouTube-like financial model (advertising revenue) will provide a solution at virtually no cost to the user.

14.6 Observations

This is a real-time application, and all such applications require high bandwidth. All the suppliers of the VCS specifically communicate to users that

"Make sure that you and the other clients have a good internet bandwidth connection." The introduction of 6G and beyond to achieve higher and higher bandwidth will continue on the path of exponential growth following Gordan Moore's law [21]. This exponential growth will inherently create a lag in implementing the latest solutions for higher bandwidth in various countries due to financial and logistics constraints of geographical area. This application will also positively impact EC and SDN. The solution given in this chapter will be a pain killer to address the difference in available bandwidth between clients.

Efficiency definition: according to various studies and reports, 11 million meetings are held daily in the United States. Low bandwidth issues impact 10% or more video conference calls, leading to problems such as poor video quality, audio delays, frozen screens, or dropped connections. The low bandwidth of a presenter can affect clients with higher bandwidth in 1.1 million video conference calls every day.

On average, organizations spend 15% of their time in meetings. If 10% of the meetings are ineffective due to low bandwidth issues, that translates to a loss of over 528,000 manhours/day (assuming an average of four persons are in the meeting) just in the United States.

The efficiency will be measured on the QoE and QoS based on the measurement of delay, jitter, and packet loss by monitoring before and after the implementation of the solution as defined by this research paper.

14.7 Conclusions and Future Scope

In conclusion, this chapter has identified the pain points in the video conference meetings and proposed a solution to cure the pain for the current application and all real-time applications, including EC and SDN. The clients with higher bandwidth would be operating at full efficiency.

The future work will complete a proof of concept and define the measurement metrics, showing the real analysis and comparison between the measures and solutions proposed in this chapter. The clients with lower bandwidth will be able to extend the life of their communication hardware infrastructure.

We will expand this solution to big data transfer applications. This research will have broader implications in remote learning, IoT, health, telemedicine, food, and lifestyle, impacting them positively.

There will be a cost to the client for cloud computing. However, for the majority, the service could be free of charge, and the cost would be offset by advertising revenue. For other clients, a subscription-based model could be used.

Acknowledgments

The authors acknowledge Raina Arora, from The Harker School, and Riya Arora, from Babson College, Boston, for their assistance.

References

[1] https://www.cengn.ca/information-centre/innovation/timeline-from-1g-to-5g-a-brief-history-on-cell-phones/
[2] K. Buchholz, "Where 5G Technology Has Been Deployed," July 27, 2022. https://www.statista.com/chart/23194/5g-networks-deployment-world-map/
[3] https://faisalchughtai.com/media/resources/Generations_of_Cellular_Communication_p4L85mi.pdf
[4] https://www.lifewire.com/6g-wireless-4685524
[5] https://iot-analytics.com/number-connected-iot-devices/
[6] https://merakiresources.cisco.com/wi-fi-6-whitepaper
[7] https://www.statista.com/chart/32691/average-number-of-devices-and-connections-per-capita
[8] A. Solyman and K. Yahya, "Evolution of Wireless Communication Networks: From 1G to 6G and Future Perspective," International Journal of Electrical and Computer Engineering, vol. 12, no. 4, pp. 3943–3950, August 2022.
[9] https://futuretimeline.net/data-trends/2050-future-internet-speed-predictions.htm
[10] https://www.rcrwireless.com/20220503/5g/5g-by-the-numbers-72-countries-nearly-2000-cities-24-standalone-networks
[11] https://typito.com/blog/video-resolutions/
[12] www.worldometers.info/geography/how-many-countries-are-there-in-the-world/
[13] https://en.wikipedia.org/wiki/List_of_countries_by_4G_LTE_penetration
[14] https://www.satelliteinternet.com/providers/starlink/
[15] https://neomeet.fi/en/blog/how-can-bandwidth-affect-video-conferencing/

[16] K. MacMillan, T. Mangla, J. Saxon, and N. Feamster, “Measuring The Performance and Network Utilization of Popular Video Applications,” Proc. of The 21st ACM Internet Measurement Conference, pp.229–244, Nov. 2, 2021.

[17] S. Afzal, V. Testoni, C. E. Rothenberg, P. Kolan, and I. Bouazizi, “A Holistic Survey of Multipath Wireless Video Streaming,” Journal of Network and Computer Applications, vol. 212, March 2023.

[18] Z. Yan and C. Chen, “Cloud-based Video Communication and Networking: An Architectural Overview,” Journal of Communications and Information Networks, vol. 1, pp. 27-43, April 2017.

[19] M. Hamidou Harouna Omar, Kouraogo Justin P., Kabre Windmi Jonathan M., Tapsoba Stanislas David Wendkouni, Sie Oumarou Pr, “Using QoE Metric as a Decision Criterion in Multimedia Heterogeneous Network Optimization: Challenges and Research Perspectives,” Journal of Computer and Communications, vol. 2024, no. 1, 2024.

[20] https://www.microsoft.com/en-us/microsoft-365/business-insights-ideas/resources/10-reasons-to-use-video-conferencing

[21] https://ourworldindata.org/moores-law

Index

About the Editors

Chih-Cheng Tseng received his Ph.D. degree from the Graduate Institute of Communications Engineering, National Taiwan University, Taipei, Taiwan, Republic of China. He is currently a full professor at the Department of Computer Science and Engineering, National Taiwan Ocean University, Taiwan. He has been actively involved in professional activities, e.g., serving as the Operations Co-Chair of IEEE GLOBECOM 2025, TPC Co-Chair of WPMC 2023 and 2024, and Sponsorship Co-Chair of IEEE GLOBECOM 2020. He is currently serving as the associate editor of the international journals, IET Electronic Letters (SCI) and Journal of Mobile Multimedia. He also served as the associate editor of Physical Communications (SCI) and the guest editor of SCI journals such as the International Journal of Sensor Networks, Wireless Personal Communications, and Wireless Communications and Mobile Computing. His research interests include the design and performance evaluation of next-generation mobile communication technologies and O-RAN-based intelligent designs.

Albena Mihovska holds a PhD degree in EEng (2008) from Aalborg University (AAU), Aalborg, Denmark. She has a strong academic research track record of more than 20 years, with positions as an Associate Professor at the Dept. of Electronic Systems at AAU (until 2017), Denmark and later as an Associate Professor in Digital Technologies at the Dept. of Business Development and Technology at Aarhus University (AU), Denmark. Currently, she holds positions as a CTO, SmartAvatar BV, Netherlands and as a Research Director at CTIF Global Capsule (CGC) Foundation, Skagen, Denmark, of which she is a Founding Member. Further, she is a Senior Research Expert at the Laboratory for Intelligent Communication and Infrastructure at Research & Development & Innovation Consortium (RDIC), Sofia, Bulgaria. She is a WG3 Vice-Chair of the one6G Association (https://one6g.org) and a Board Member of EUROMERSIVE (https://euromersive.eu), and President of the INFORMS Telecommunications & Networks Analytics Cluster. Her research interests are in 6G connectivity solutions for applications across vertical market segments. Dr. Mihovska has a very strong track record as a Project Coordinator, Technical Manager and Team Leader of various size EU-funded projects. She is appointed as IEEE R8 Standards Coordination Committee Member 2025 and is the external liaison of one6G to IEEE P1955 standardisation group. She serves on the Steering Committee of IEEE Black Sea Communications Conference (BLACKSEACOM) since 2015.

Shao-Yu Lien received his Ph.D degree from the Institute of Communication Engineering, National Taiwan University. He is now a professor of the Institute of Intelligent Systems, National Yang Ming Chiao Tung University (NYCU), Taiwan. Dr. Lien is also a technical director of Institute for Information Industry (III), Taiwan, since 2020. Before joining NYCU, Dr. Lien was an associate professor of the Department of Computer Science and Information Engineering, National Chung Cheng University (CCU), Taiwan. Dr. Lien has received a number of prestigious research recognition, including IEEE ICC 2010 Best Paper Award, IEEE Communications Society Asia-Pacific Outstanding Paper Award 2014, WPMC 2022 Best Paper Award, IEEE PIMRC 2023 Best Student Paper Award, IEEE APWiMob 2023 Best Paper Award. Dr. Lien is also a World's Top 2% Researcher in 2020-2024. He also served as the guest editor of IEEE Transactions on Cognitive Communications and Networking (TCCN), IEEE Transactions on Vehicular Technology (TVT), IEEE Transactions on Green Communications and Networking (TGCN), and Wireless Communications and Mobile Computing (WCMC). In the meantime, Dr. Lien also served as the leading organizers of a number of technical workshops in IEEE VTC-Spring 2015, IEEE GLOBECOM 2015, Qshine 2015 and 2016, IEEE PIMRC 2017, IEEE GLOBECOM 2019, and IEEE ICC 2020. Particularly, he has been a 3GPP standardization delegate since 2009 for LTE, LTE-A, LTE Pro, 5G NR, and 5G-advanced, and in this role, he has contributed more than 70 technical documents and patents. Dr. Lien's research interests include AI/ML computing, 6G mobile networks, and robotic networks.

For Product Safety Concerns and Information please contact our
EU representative GPSR@taylorandfrancis.com Taylor & Francis
Verlag GmbH, Kaufingerstraße 24, 80331 München, Germany